Space and Society

The Space and Society series explores a broad range of topics in astronomy and the space sciences from the perspectives of the social sciences, humanities, and the arts. As humankind gains an increasingly sophisticated understanding of the structure and evolution of the universe, critical issues arise about the societal implications of this new knowledge. Similarly, as we conduct ever more ambitious missions into space, questions arise about the meaning and significance of our exploration of the solar system and beyond. These and related issues are addressed in books published in this series. Our authors and contributors include scholars from disciplines including but not limited to anthropology, architecture, art, environmental studies, ethics, history, law, literature, philosophy, psychology, religious studies, and sociology. To foster a constructive dialogue between these researchers and the scientists and engineers who seek to understand and explore humankind's cosmic context, the Space and Society series publishes work that is relevant to those engaged in astronomy and the space sciences, while also being of interest to scholars from the author's primary discipline. For example, a book on the anthropology of space exploration in this series benefits individuals and organizations responsible for space missions, while also providing insights of interest to anthropologists. The monographs and edited volumes in the series are academic works that target interdisciplinary professional or scholarly audiences. Space enthusiasts with basic background knowledge will also find works accessible to them.

More information about this series at http://www.springer.com/series/11929

Sandra Häuplik-Meusburger • Sheryl Bishop

Space Habitats and Habitability

Designing for Isolated and Confined Environments on Earth and in Space

Springer

Sandra Häuplik-Meusburger
Institute of Architecture and Design
TU Wien
Vienna, Austria

Sheryl Bishop
University of Texas Medical Branch
Galveston, TX, USA

ISSN 2199-3882 ISSN 2199-3890 (electronic)
Space and Society
ISBN 978-3-030-69739-6 ISBN 978-3-030-69740-2 (eBook)
https://doi.org/10.1007/978-3-030-69740-2

Cover design: Paul Duffield

This Springer imprint is published by the registered company Springer Nature Switzerland AG.
The registered company address is: Gewerbestrasse 11, 6330 Cham, Switzerland

Foreword

When I was a young boy, and my mother wanted to get the truth from me, she would capture my attention and say; "Tell me what *really* happened."

This book is like that: It captures your attention, and lets you look long and honestly at what's happening as we gain experience for projected long-term living in space. It surveys the many earth-bound "analogue" studies done, and summarizes the on-orbit experience to date, including the former Soviet Salyut and Mir, the US Skylab, and current International Space Stations. Across all of these, the lessons are clear: We can keep space crews alive and working productively in the confines of space craft, but we have not yet learned how to give them a *home away from home.* Living in space has been until now a high-tech camping trip at best. As we extend stays on the Moon and possibly Mars, it is going to have to become something different, and more. Habitats will have to become supportive and nurturing homes for their inhabitants.

The authors are well suited to make this case: One is a Social Psychologist with decades of experience in the Extreme Environments field. The other is a Space Architect with the keen sense of how Design plays a crucial role in how crews of Isolated and Confined Environments experience and perform throughout their stays. Together, they show through the interviews they've collected that while we've got the engineering right for the most part, we are just beginning to appreciate the social psychology involved, and we as yet have to acknowledge effects of the *habitat design.*

Hopefully, their book will help change all that. It is really something that every spacecraft engineer and planner should read: This is what is actually happening to the people we are sending aloft. It is a testament to their bravery and fortitude that they have performed as well and consistently as they have. But it has not been because of the design of what they lived in, but mostly in spite of it. And the communications protocols, the command structures, the crew selection procedures have not helped all that much either, and have occasionally proved almost disastrous. But crews find workarounds, and endure, and occasionally rebel. But they do not *thrive*. And that is what they will have to do if we envision years or more far away from Earth.

It is tempting to think that all of this is a question of improved technology: Give them more video games, higher fidelity communications, a space "smartphone," perhaps. But it's not. It's about how they live and interact every day, and how the environmental design of the habitat supports all of that. And what this book brings clearly home is that we haven't begun to think about how good that could be. We've taken the "bargain basement" approach instead, thinking that if we get the physics and the basic physiology supported, all will be well. But it won't be, and this book shows why.

Reading it is in some ways like sitting down in a quiet "Bier Stube" with all of those former crewmembers of all the analogue bases and previous space stations and asking them: "Tell me what really happened." Then listening very closely, because through the pages of this book, they will tell you.

Richland, WA, USA James A. Wise

Preface

Objective of the Book

Human factors and habitability are important topics for working and living spaces. For space exploration, they are vital for mission success. Human factors and certain habitability issues have been integrated into the design process of human operated spacecraft; however, there is a crucial need to move from mere survivability to factors that support thriving. 'Habitability' and human factors will become even more important determinants for the design of future non-commercial and commercial spacecraft and extraterrestrial habitats as larger and more diverse groups occupy off-earth habitats for longer periods of time. The '*risk of an incompatible vehicle or habitat design*' (NASA [HSIA], 2020)[1] has been identified by NASA as a recognized key risk to human health and performance in space.

This book provides an overview of the historic advancements of human operated spacecraft, as well as highlighting various current and future concepts of *habitability* and their translation into design. The main goal of this book is to promote a dialogue between the diverse concepts of *habitability* and their socio-spatial and psychological dimensions. Selected dimensions are reviewed from multiple backgrounds, and possible design and architectural related applications are illustrated and discussed. The authors explore various concepts of the term *habitability* from the perspectives of the inhabitants as well as the planners and social sciences and highlight common features and differences. A major focus of this book is to explore and hopefully stimulate creative solutions to the unique challenges inherent in crafting livable spaces in extraterrestrial environments, fostering a constructive dialogue between the researchers and planners of future inhabited spacecraft and extraterrestrial habitats.

[1]NASA [HSIA], 2020. Risk of Adverse Outcome Due to Inadequate Human Systems Integration Architecture, published 07/30/20. https://humanresearchroadmap.nasa.gov/risks

The book will benefit individuals and organizations responsible for human operated space missions. It also provides insights of interest to researchers of social sciences, psychology, engineering, architecture, and design.

The focus of this book is on the issue of designing living spaces for extreme environments and, particularly, extraterrestrial environments. Its findings address the basic socio-spatial relationships as they are applicable to other extreme environments and even for compact urban living on Earth. In addition, it unveils the authors' experiences of the relationship between habitable space and the environment and thus relates back to Earth and our very normal daily life.

How This Book Came into the World

> Psychologists and architects are natural allies, joined by their search to support people in their varied endeavors[2] (Harrison 2009, p. 890).

The authors share a common passion on the relationship between space and humans, and have approached their research from different angles. While Sheryl has always been hunting for the answer to the question of '*what constitutes the "best" fit person to live happily in extraterrestrial environments*', Sandra has been chasing after '*what constitutes the space to live happily in (extraterrestrial) environments*'. This book is a real synergetic achievement towards designing living spaces for space environments, and this chapter tells you how it started.

Sandra Häuplik-Meusburger and Sheryl Bishop have known each other for many years through their work. They first met in person at a conference in Glasgow in 2008. A bit later they started to collaborate on research projects and publications. And during a lunch in Vienna in 2017, collaboration for a book project was set into motion. Before the authors talk about their collaborative work on this project, read the individual stories of Sandra and Sheryl.

Sheryl's Story

I can remember the sense of relief I felt when I discovered Bertalanffy's *General System Theory* (1968)[3] in college which defined a system as a complex of interacting elements, open to, and interacting with, their environment, acquiring qualitatively new properties through emergence and continually evolving. Bertalanffy considered general system theory a 'general science of wholeness'. In an era where

[2]Harrison, A. A. (2010). Humanizing outer space: architecture, habitability, and behavioural health. *Acta Astronautica 66*(5–6), 890–896.

[3]Bertalanffy, L. (1968). General System Theory: Foundations, Development, Applications. New York: George Braziller.

reductionism and specialization was king, I thought '*Finally, something that fits me*'. In his advocacy of interdisciplinary inclusiveness, I found permission to pursue a career as a 'generalist' instead of following the trend of professional specialization. As a social psychologist, I was free to pursue all aspects of human development across the entire lifespan and range across the vast spectrum of 'normal' (aka non-pathological) human behavior as my interests took me. It served me well as I pursued the fairly esoteric answer to 'what constitutes the 'best' fit person to live happily in extraterrestrial environments'? Answering THAT question would lead me to pursue hundreds of factors that influence wellbeing and afforded me opportunities to work with all kinds of engineers, architects, biologists, neurologists, geologists, physicians, astronomers, astrophysicists, chemists, pharmacists, and not a few of my own colleagues from the fields of psychology and psychiatry over the last 30 years.

In the mid-1980s, I had the good fortune to stumble across the work of a fellow psychologist/mathematician/environmental scientist, Jim Wise, who was proposing that including patterns in interior designs that mimicked the same kinds of patterns found in nature improved human performance and wellbeing. His work and subsequent others (e.g. Wise and Rosenberg 1986; Wise and Taylor, 2002)[4] suggested that we could craft human habitats that were automatically functionally beneficial and supportive in addition to providing shelter and ensuring survival. Coupled with the myriad evidence for the beneficial effects of including 'natural' elements (aka plants, trees, water features) in the 'built environment' from other sources, there was an explosion of interest and research on incorporating nature, eventually called biophilic design, into every aspect of human habitation. And, thus, my research into how to ensure optimal adaptation and functioning for small groups situated in extreme environments such as those our future space explorers would face intersected with the world of architecture and interior design. Because *where* we live affects *how* we live.

The early attempts to explore how incorporating natural elements affected individual and group dynamics were met with large indifference from the primarily engineering-oriented research community associated with space. Humans were viewed as those irritatingly inconsistent and unpredictable 'users' that repeatedly went 'off-nominal' with virtually everything involved in a spacecraft. They demanded windows to look out of when there was no functional need for windows, cues for 'up' and 'down' in a microgravity environment that could take advantage of 360° of surface and unconsciously circumvented surfaces used for eating instead of taking the more efficient path of floating over them. There were irreconcilable cultural differences over the types of hygiene facilities desired (showers versus saunas) and water treatment (iodine versus biocidal silver) that resulted in duplicated

---

[4]Wise, J. A. & Rosenberg, E. 1986. *The effects of interior treatments on performance stress in three types of mental tasks*. Technical Report, Space Human Factors Office, NASA-ARC, Sunnyvale CA.; Wise, J. A., Taylor, R. P. (2002). Fractal design strategies for enhancement of knowledge work environments. *Proceedings of the Human Factors and Ergonomics Society Annual Meeting, 46*(9), 854-858.

systems (water treatment) and initially minimal bathing facilities (washcloths) for the International Space Station. Proposals to add such unessential aesthetics such as color, textiles, or textures much less actually introduce living plants were enough to put architects and environmental/social psychologists on the 'do-not-invite' list for space project planners. But we simply would not go away. As my circle of architect colleagues expanded and I was blessed with opportunities to work with individuals like my co-author, the need to move to 'thriving not just surviving' became undeniable and eventually was incorporated into NASA's Human Research Roadmap. As with all paradigm shifts, rhetoric eventually leads to action. All I can say is that it was about time!

Sandra's Story

I love the combination of science and fiction, in that it includes a vision of where we want to go and a possible path to reach it. I have always wanted to know how things work and how they are connected. When I started to study architecture, I became even more intrigued by the interrelations between space and humans. I had always felt that built space influences human behavior and now I was about to look behind the mechanics of this 'human-space system'.

I enjoyed lessons on how material properties can change user behavior and that the actual physical surroundings are related to certain human behavior. Humans read space; they connect physical appearance to function, to social behavior. Without having to think about it we (in the Western hemisphere) 'know' how to behave in a church, in a dwelling area, or at a playground. No signs are necessary. This close relationship between *built space* and *lived space* slowly became unveiled and I started to translate those experiences into real architecture. My first designed and built space was a fashion shop in Vienna for one of my best friends. It was a tiny space, and had to function as a tailor's workshop, a showroom, as well as an event location. I designed every single piece of interior, from the lamp to the stowage space, and we choose the materials very carefully. The interior concept followed the idea of a catwalk and became a multipurpose yet minimalistic space. Today in 2020, my friend is one of the top Austrian designers and the shop interior still fits the changing garments and activities. Retrospectively, I also realized that in order to create a real sustainable space, it is of huge importance to identify (which is a most complex and sensitive task) and integrate the client's requirements and identity into the design process from the very beginning.

Later, for my dissertation I chose to research the relationship between humans and 'space' (the social and physical space they live in) in an extreme environment. The idea was that in an *extreme environment* (see Chap. 1), where inhabitants cannot exist without a built environment, where they heavily rely on 'what is there', and where they 'cannot go outside to take a breath of fresh air', I would discover something new.

I was blessed to have Richard Horden, former professor in Architecture and Product Design at the Technical University of Munich[5] and the creator of the term 'micro architecture',[6] as my dissertation mentor. He was a wonderful person to talk with and I still have great appreciation for his way of looking at the world. He had a minimalistic approach to architecture, in that the 'space in-between' was more important for him than the structure. He wanted to only 'touch the Earth lightly' with his projects. In addition, he introduced me to the world of product and industrial design.

My PhD resulted in a comparison and analysis of all inhabited human spacecraft and space stations in relation to human activities that later turned into the book *Architecture for Astronauts*.[7] I evaluated those extraterrestrial habitats from a human activity point of view: The Apollo Spacecraft and Lunar Module, the Space Shuttle Orbiter, and the Space Stations Salyut, Skylab, Mir, as well as the International Space Station. To facilitate orientation and to ease comparison with architectural drawings and diagrams, each category was assigned a specific color. Design directions for each category concluded each chapter. Next to an overview of the architecture and configuration concerning the interior layout and a comparison of the spatial and time allocation of human activities, the main part of the book concentrated on the investigation of the relationship between the environment and its users. At that time, it was the first compendium that included comparable information of all inhabited spacecraft from an architectural and user point of view. For that, I have to thank the many astronauts and cosmonauts that I personally spoke to. Their unique input became very valuable, even trailblazing, and often contradicted some official views and revealed something new.

The search for '*what constitutes the space to live happily in (extraterrestrial) environments*' is continuing, and I am grateful to integrate my research with my work as an architect, on Earth as off Earth.

Towards a Habitability for Humanity

This book is a real synergetic achievement towards designing living spaces for extraterrestrial environments, and this chapter tells you how it started. It started with the idea of writing a compendium on space habitability from the human point of

[5]Fusing high-tech engineering with industrial-design methods, he and his research students in Munich have created an innovative range of revolutionary buildings in a broad variety of settings. From the Ski Haus (delivered to the Alps by helicopter and used by mountaineering and rescue teams) and Antarctic living modules to the Micro Compact home, a fully self-contained pre-fab home that fits into a 2.65 m^2 cube, these structures are designed for their adaptability to our changing planet, lifestyles, and basic human needs.

[6]See the book: Micro Architecture: Lightweight, Mobile and Ecological Buildings for the Future by Richard Horden, 2008, Thames & Hudson.

[7]Häuplik-Meusburger, S. (2011). *Architecture for Astronauts – An Activity based Approach.* Springer Praxis Books, Vienna.

view, and became elaborated, queried, and adapted with many discussions in both personal and online meetings.

In the middle of the book process we decided to do a habitability survey and integrate it into a new chapter. The work process was intensive as we really worked in an interdisciplinary way. The questionnaire was developed together, discussing all the questions. The list of possible participants was gleaned from both of our extensive networks of acquaintances, friends, colleagues, collaborators, and all their contacts. Distributed by word of mouth and through emails and social media, we received responses from around the world from those that had spent time in isolated, confined environments both here on Earth and aloft. The scope of the response was an amazing testament to how interconnected we all are: dozens of professions, ages that ranged from 29 to 72, and participants from 15 different countries. Each gave us the benefit of their experience. It took our interdisciplinary collaboration to new heights. Instead of just two of us, we became many with their willingness to share. That set the tone of the entire book. We strove to elaborate on almost everything with personal quotes from others. Instead of confining ourselves to a preplanned outline, we followed the flow of emergent issues. We did not cover everything there is to cover . . . the book would have been five times longer. We may not have even covered all the really important things. As you will see, what is important frequently differs. But we hope that we have covered *enough* to excite discussion and further exploration among the worldwide community.

Vienna, Austria — Sandra Häuplik-Meusburger
Santa Fe, TX — Sheryl Bishop

Acknowledgements

This is the most difficult chapter of all. There are so many people to acknowledge on our paths. Some gave of their personal experience and insights in analogue environments; others provided feedback on the chapters, pictures from their own missions, assistance in procuring permissions to use other photos, or suggestions for additional content. To all those who participated either through our survey or as unsung supporters, content experts, and hands-on editors, we thank you. Special thanks to Executive Editor Ramon Khanna and Editorial Assistant Rebecca Sauter for their invaluable guidance throughout the publication process; Donald C. Baker for his assistance with the design of our analogue experience survey; and Irina Panturu for updating and creating some of the graphics in this book. Deepest gratitude to Zita Barcza-Szabo for her hard work and long hours which contributed to the final editing phase prior to publication. To all that walked beside us on this journey, the list of acknowledgements below is not complete by any means. We would like to especially thank the following people (in alphabetical order) as well as those who chose to remain anonymous for taking part in our habitability survey.[8]

Clemente Balladares, Olga Bannova, Tristan Bassingthwaighte, Donald Barker, K Bhattacharyya, Kristian von Bengtson, Koray Bingol, Anne Burley, Doug Campbell, Bill Clancey, Jean-Francois Clervoy, William Daniels, Brian Dykas, Timothy Evans, Marinella Ferrino, Kyle Foster, Ron Franco, Rahul Goel, Ethan Good, Alice Gorman, Brand Griffin, Gernot Groemer, Patrick Harkness, Zan Hammerton, Chiemi Heil, Martha Henriques, Carmel Johnston, Jesper Jorgensen, Claudia Kessler, Taber MacCallum, Julia DeMarines, Kenny Mitchell, Ana Mosquera, Euan Monaghan, Brent Monseur, Guy Murphy, Nicolas Nelson, Carrie Paterson, Maria Antonietta Perino, Vladimir Pletser, Dumitru-Dorin Prunariu, Linda Roehrborn, Jackelynne Silva-Martinez, Jonna Ocampo, Shayan Shirshekar, Barret Schlegelmilch, Jack W. Stokes Jr., Eirik Sønneland, Peter Suedfeld, Daniel Surber,

[8]The list is not complete as some of the interviewees preferred to answer anonymously.

Rachel Tillman, James Titus, Miha Tursic, Truman Young, Rochelle Velho, and Cyprien Verseux.

Special acknowledgement goes to James Wise and Georgi Petrov for their thorough review of the final manuscript.

Finally, and perhaps most important, the authors are grateful to their families, especially their husbands and children, for their patience and emotional support while working on this book.

Contents

About the Authors

Sandra Häuplik-Meusburger PhD, is an architect and researcher specializing in compact habitability design solutions for extreme environments. She teaches at the Institute for Architecture and Design at the Vienna University of Technology and is appointed academic director of the course Space at the Science Academy in Lower Austria. Sandra uses cross-program comparison and analysis of inhabited isolated, confined, and extreme environments (ICEs) on Earth and space from a human perspective as a basis for the systematic assessment of existing and future living and working environments. She is corresponding member of the International Academy of Astronautics (IAA) and Vice-chair of the Space Architecture Technical Committee (SATC) of the American Institute of Aeronautics and Astronautics. Sandra has published several scientific papers and is author of the books *Architecture for Astronauts* (Springer, 2011) and *Space Architecture Education for Engineers and Architects* (Springer, 2016).

Sheryl Bishop PhD, is a social psychologist and behavioral scientist who has applied her skills widely over her career. A Professor Emeritus at the University of Texas Medical Branch, she has taught in all four schools, Medicine, Nursing, Health Sciences, and the Graduate School of Biomedical Sciences as well as served as faculty for the International Space University, Strasbourg, France. As an internationally recognized behavioral researcher in extreme environments, for the last 35 years Dr. Bishop has investigated human performance and group dynamics in teams in extreme, unusual environments, involving deep cavers, mountain climbers, desert survival groups, polar expeditioners, Antarctic winter-over groups, and various simulations of isolated, confined environments for space, including a number of missions at remote habitats. She has routinely presented her research at numerous scientific conferences, has over 60 publications (including contribution to NASA's Historical Series on Psychology in Space), and over 50 scholarly presentations in both the medical and psychological fields. She is frequently sought out as a content expert by various media and has participated in several television documentaries on space and extreme environments by Discovery Channel, BBC, and 60 Minutes.

Chapter 1
Introduction

This chapter outlines why an extraterrestrial environment can be considered an Extreme Environment (EE) and how extreme environments entail demands that cannot be ignored or underestimated. Habitability considerations for such conditions are vastly impacted. A discussion of how the general envelope of human factors and habitability are affected by the unique characteristics of Isolated and Confined Extreme Environments (ICEs) is presented with an emphasis on the costs of not including habitability as a critical design component. Basic assumptions, the selection of examples used throughout the book and sources of information are identified.

1.1 The Unforgiving Environment

When we talk about the environment of 'outer space' or **extraterrestrial environments**, we refer to a natural environment that is beyond Earth or not from Earth. According to our current knowledge there is no extraterrestrial environment that is naturally livable for human beings as they lack critical resources (e.g., breathable air, water) or involve hostile environmental challenges (e.g., microgravity, high radiation). As such extraterrestrial environments are categorized as **extreme environments (EE)**.

Any environment can be described through their environmental or physical variables, such as temperature, lighting and specific characteristics, and also through their psychological and social variables. Psychologist Peter Suedfeld describes *extreme (or unusual) environments* in contrast to 'normal environments' along two dimensions: (1) degree of extremeness and (2) unusualness (Kring 2008, p. 212; Suedfeld and Mocellin 1987). As such, an extraterrestrial environment would be considered one of the most extreme environments with a high degree of extremeness (physical dangers and discomfort) and a high degree of unusualness as contrasted to a normal environment. Psychologists Manzey and Lorenz (1997) defined EEs as

S. Häuplik-Meusburger, S. Bishop, *Space Habitats and Habitability*, Space and Society, https://doi.org/10.1007/978-3-030-69740-2_1

"*settings for which humans are not naturally suited and which demand complex processes of psychological and physiological adaptation*".

EEs are commonly referred to as *settings that possess* ***extraordinary (extreme and unusual) physical, psychological, social (or interpersonal), and technological demands*** *that require significant human adaptation for survival and performance* (Suedfeld and Mocellin 1987; Manzey and Lorenz 1998; Kring 2008) (see Box 1.1).

Box 1.1 Extreme Environments: Definition

Jason Kring (Kring 2008, p. 212) has summarized the findings on EEs as follows:

First, extreme environments possess extraordinary and unique features in three key areas:

- Physical characteristics of the environment,
- Interpersonal and social dynamic of individuals and groups living and working together in the environment,
- Psychological variables that influence how an individual responds to the environment.

Second, EEs require significant human adaptation in order for people to live and work in a manner that supports physical and psychological health, promotes successful task performance, and protects individuals from injury (Kring 2008, p. 212).

The social, psychological and also spatial significance of living in an extraterrestrial environment has become more explicitly characterized by a further refinement of the term to incorporate the extraordinary isolation and confinement found in all extreme environments. Such environments are those in which "*physical parameters [...] are [...] outside the optimal range for human survival [...] and which conditions [...] deviate seriously from the accustomed milieu of most [and further] involve physical remoteness [...] and a circumscribed spatial range*" (Suedfeld and Steel 2000, p. 228).

Human challenges of ICEs include: prolonged isolation and confinement, a hostile natural environment with limited mobility outside the habitat, high autonomy of the crew and habitat, life in a 'microsociety'[1] (and often in a multi-cultural setting), and a high probability of under-stimulation and boredom on long-duration missions, etc. Such conditions place demands not only on the type of persons who would be 'best fit' to inhabit such environments but also on the living spaces that must be crafted to support human habitation in such environments. Next to extra-terrestrial habitats, polar and underwater habitats are examples of this building typology.

[1]A small, self-contained community or milieu governed by its own conventions, rules, etc.

Table 1.1 Physical and social factors that have been reported to cause stress (Evans et al. 1988, p. 4; Connors et al. 1985; Cohen and Häuplik-Meusburger 2015)

Physical stressors associated with ICEs	Social and Psychological stressors associated with ICEs
Changes in pressure Extreme temperatures Unusual environmental hazards (meteorites, radiation, etc.) Physical threat to life in exterior environment Loss or alteration of time markers/ Zeitgeber Irregular or unnatural light cycles Noise and vibrations Limited available space Short focal distance stressed near-sightedness Poor ventilation Sterile and monotonous surroundings Restricted diet	Isolation and confinement The feeling of being crowded The feeling of loneliness and separation from one's normal social group Reduction of privacy The necessity of forced interaction with a small group of people Dependence on a limited community Disconnection from the natural world No separation of work and social life No family life Repetitive and often meaningless tasks Limited habitability (limited hygiene, sleep facilities, isolation from support systems)

One of the critical characteristics for living and working in those environments—and thus mission success—is the dependency on the habitat, its technological capability as well as the sociospatial framing. Inhabitants, who are exposed to remote and hostile environments, not only must overcome the challenges posed by the dangers and limitations imposed by the particular environment itself, but also experience significant distress from being confined indoors and isolated from civilization and social contact. Table 1.1 lists a selection of physical and social factors that can become stressors for the inhabitants.

All of these factors and their associated stress responses must be taken into consideration when designing livable space or habitats for ICE environments. Yet, historically, such habitats have lacked all but the merest attention to such details with a focus primarily on surviving rather than thriving. This is changing and the built environment is slowly becoming an important factor to ensure both physical and psychological wellbeing.

1.2 Defining Habitability

"*Early spacecraft had been designed to be operated, not lived in*" (Compton and Benson 1983, p. 130). Mercury[2] astronauts did not climb into the spacecraft, they put it on. At that time, spacecraft design was primarily functional. After the first space

[2]Project Mercury was the first human spaceflight program of the United States and ran from 1958 to 1963.

missions, when NASA and the Soviets were advancing their goals for long duration missions to prove that humans could live and work in space for extended periods, the habitable design of the interior became increasingly important.

In 1963 Reed and White started their paper with the question "*What does habitability mean?*" (Reed and White 1963). Interestingly, their examination of the meaning directly led to the discussion of the interplaying factors "*related to the man, **his** machine, **his** environment and the mission*". Kubis (1967, p. 399) stated a few years later that habitability "*depends upon the purpose of man's presence within that environment, the time he plans to remain, and the type of performance he expects to achieve*". Psychologist Dr. James Wise defined **spatial habitability** as: "*Spatial habitability refers to the ways in which the volume and geometry of livable space affects human performance, wellbeing and behaviour*" (Wise 1988, p. 6).

Today, the term, **habitability,** is understood to be an umbrella term that describes the suitability and value of a built habitat for its inhabitants in a specific environment (cf. Häuplik-Meusburger 2011). It is a complex system related to the individual as well as society in relation to the (built) environment. Figure 1.1 highlights the interrelating factors between the inhabitant(s) and the lived-in environment.

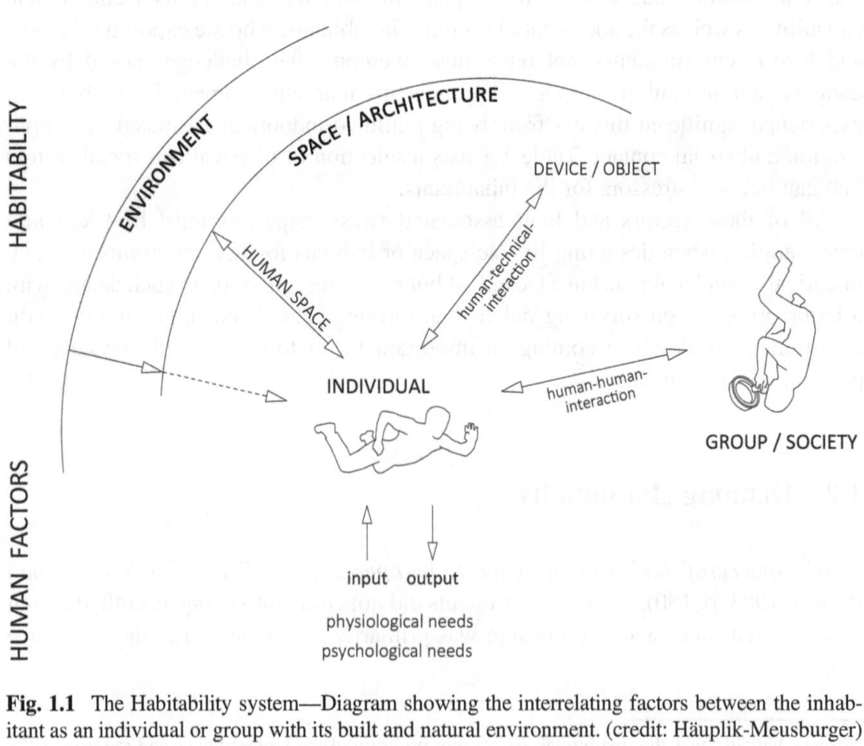

Fig. 1.1 The Habitability system—Diagram showing the interrelating factors between the inhabitant as an individual or group with its built and natural environment. (credit: Häuplik-Meusburger)

1.2.1 Elements of Habitability

Components of the system (Fig. 1.1) include:

The **Setting**: The physical environment in which human operated missions take place is life-threating and physically, psychologically and socially demanding. The longer the mission lasts the more strenuous it is psychologically for the individual crewmember, as well as for the whole crew. In addition, longer missions place more sophisticated technological demands on the habitat and associated technology and systems. Subcomponents include conditions of the actual environment, mission length, tasks, type of habitat, and others.

The **Individual**: So far individuals for space missions come from a small spectrum of our society. They have been carefully selected based upon specific characteristics, based on either select-out (physical and/or psychological disabilities) or select-in criteria (knowledge, experience, personality, etc.). Subcomponents include physical and psychological condition, behavioral health, experience, and others.

The **Group or (Micro)society**: In extraterrestrial habitats a small group of people are living together in a relatively small space. This is often referred to as a micro-society. Isolated from the normal social matrix on Earth, social relationships often become more intense and, thus, can produce interpersonal challenges for the whole group. Subcomponents include: crew composition, selection, gender, culture, and others.

The **Time**: The length of the mission affects every component—the individual, the whole group—as well as the habitat and technical facilities. Subcomponents include: mission length, changes during the mission, and scheduling.

1.3 The Cost of (Not Including) Habitability

How much does it cost to include habitability into the design process? And how much will it cost NOT to include habitability into the design process?

In the 1963 Manned Space Laboratory Conference, Celentano, Morelli and Freeman emphasized the importance of habitability with an excerpt on 'Habitability problems aboard Navy vessels' from 1891. "*The physical condition of the men, when it comes to action or to conditions of war, is of greater moment than the … extra knots … for which we are asked to sacrifice so much*". Later in the same paper the author reminded us that "***each generation of designers**, in their zeal for getting the most with the least, and packing the greatest into the smallest (or lightest), **frequently forgets the essential factors of habitability and needs to be reminded***" (Celentano et al. 1963).

Support and evidence for the need of integrating habitability can be found in every decade. Thomas M. Fraser suggested "*that habitability can be considered as that equilibrium state, resulting from **man-machine-environment-mission***

***interactions** which permits man to maintain physiological homeostasis, adequate performance, and psycho-social integrity*" (Fraser 1968, p. V). Later Harrison (2010, p. 891) pointed out the relation between the benefits and costs. He stated that environmental stressors "*magnify risk, undermine performance, and raise the human cost of occupancy*". For him, it is clear that "*by minimizing environmental stressors*" habitability "*serves the interests of behavioral health*". Embedded into the discussion on human factors, habitability is a major factor in the design of any inhabited environment, including its facilities. However, in extraterrestrial and isolated and confined environments, habitability is critical for the physical and psychological wellbeing of the inhabitants as well as mission success. In extraterrestrial environments and most of the environments characterized as ICEs, the basic requirements of human existence can only be secured by an additional technical envelope, such as the habitat or a space suit. Isolated from the Earth, astronauts must live for long durations within a small and confined environment, completely dependent on mechanical and chemical life support systems. This building type must be subject to careful design, planning and construction that incorporate both elements for surviving as well as those for thriving.

Habitability design integration is an important aspect when planning long-duration missions. To date, given historical missions missions of short duration, it was seen as 'nice to have' but it becomes vital when mission length increases. As stated by Frances Mount (Mount 2002, p. 87): "*The impact of a poorly designed switch or lack of stowage area is different for a mission of six months compared to a mission of one week.*"

The research into effective habitat design for human health and wellbeing in isolated, confined and extreme environments is still nascent and distributed across multiple disciplines. The recent intersection of several critical lines of research are beginning to frame multiple intriguing new approaches to implementing environmental design in ways that integrate 'place and use demands' as deliberative elements fostering restoration, wholeness and maximizes adaptation and response through evolutionarily honed cognitive processes. Paired with the power of new technological tools for delivering levels of realism that are increasingly difficult to distinguish from 'reality', habitation has expanded beyond the mere physical space we occupy into the perceived space around us. We hope this book will take that discussion and exploration beyond what has been addressed before into those brave new worlds of possibilities ahead of us.

1.4 Basic Assumptions

There are a number of basic assumptions related to future extraterrestrial missions that underlie the areas dealt with in this book. These assumptions will be discussed in detail in the different chapters of the book, but it is worthwhile to mention them now as an orientation to the rest of the book.

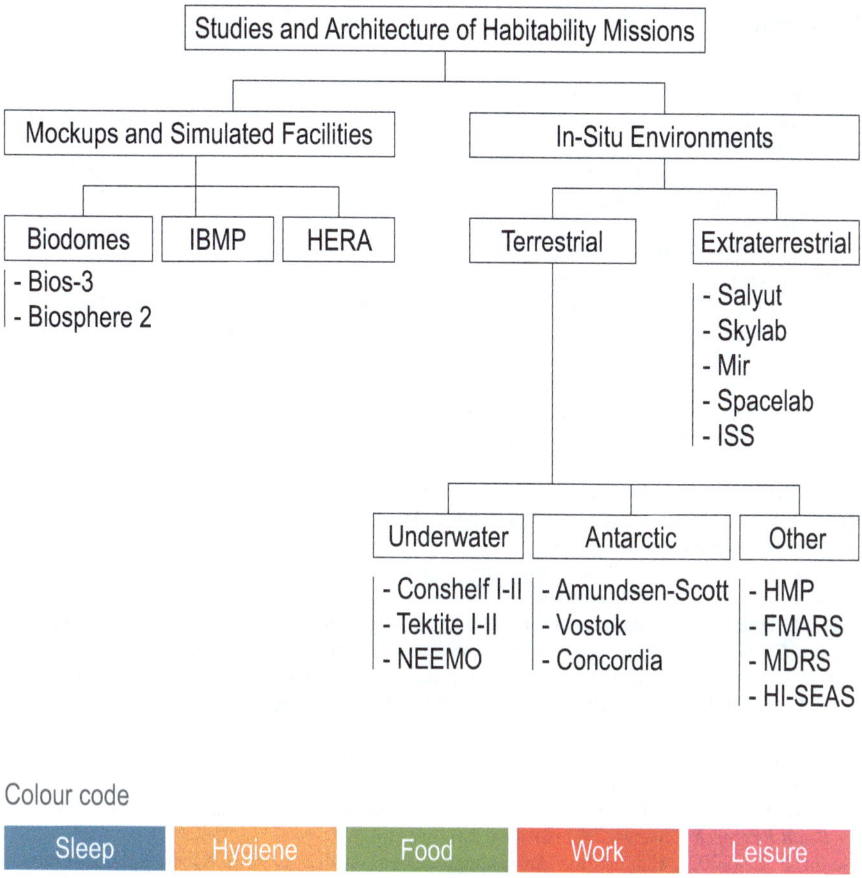

Fig. 1.2 Overview of habitability studies and relevant habitats that are introduced in Chaps. 4 and 5. A color code is used to make comparison of the plans and sections easier and comparable

Selection of examples. This book is a co-project combining the psychology of a person living in extreme environment with the socio-spatial dynamics of a group living in extreme environments. Examples are largely drawn from missions that involve living over a period of time in a habitat that serves as protection from the extreme environment (see Fig. 1.2). Expeditions, i.e., missions that involve constantly moving from one point to another with transient 'camps' along the way, are only mentioned when it is relevant to understand the context.

Sources of Information. Reports and personal experiences from inhabitants have been one of the most important sources for both authors. They are used for describing the research topic and for the definition of preliminary relevant topics. Sources include written as well as oral material from interviews by and with inhabitants of extreme environment facilities. Furthermore, a number of relevant studies from the fields of space psychology, sociology, extreme environmental medicine, as well as personal reports from analogue simulations are referenced.

References

Celentano, J., Amorelli, D., & Freeman, G. (1963). *Establishing a habitability index for space stations and planetary bases*. Retrieved May 18, 2020, from https://arc.aiaa.org/doi/abs/10.2514/6.1963-139

Cohen, M. M., & Häuplik-Meusburger, S. (2015). *What do we give up and leave behind?* ICES-2015-56, 45th international conference on environmental systems, Bellevue, Washington, 12–16 July 2015.

Compton, W. D., & Benson, C. D. (1983). *Living and working in space, a history of Skylab*. Washington, DC: National Aeronautics and Space Administration.

Connors, M. M., Harrison, A. A., & Akins, F. R. (1985). *Living aloft. Human requirements for extended spaceflight*. Washington, DC: NASA Scientific and Technical Information Branch.

Evans, G. W., Stokols, D., & Carrere, S. (1988). *Human adaptation to isolated and confined environments: Preliminary findings of a seven month Antarctic winter-over human factors study*. NASA contractor report NASA CR-177499. University of California, Irvine. Retrieved May 15, 2020, from http://www.spacearchitect.org/pubs/NASA-CR-177499.pdf

Fraser, T. M. (1968). *The intangibles of habitability during long duration space missions*. Retrieved May 18, 2020, from https://ntrs.nasa.gov/archive/nasa/casi.ntrs.nasa.gov/19680017230.pdf

Harrison, A. A. (2010). Humanizing outer space: Architecture, habitability, and behavioral health. *Acta Astronautica, 66*(5–6), 890–896.

Häuplik-Meusburger, S. (2011). *Architecture for astronauts – An activity based approach*. New York: Springer Praxis Books.

Kring, J. (2008). Human performance in extreme environments. In S. F. Davis & W. Buskist (Eds.), *21st Century psychology – A reference handbook* (pp. 210–219). Newbury Park, CA: Sage.

Kubis, J. F. (1967). Habitability: General principles and applications to space vehicles. In H. Bjurstedt (Ed.), *Proceedings of the Second International Symposium on Basic Environmental Problems of Man in Space*. Vienna: Springer.

Manzey, D., & Lorenz, B. (1997). Human performance during prolonged space flight. *Journal of Human Performance in Extreme Environments, 1*(2), 68.

Manzey, D., & Lorenz, B. (1998). Mental performance during short-term and long-term spaceflight. *Brain Research Reviews, 28*(1–2), 215–221.

Mount, F. E. (2002). Habitability: An evaluation. In H. W. Lane, R. L. Sauer, & D. L. Feeback (Eds.), *Isolation: NASA experiments in closed-environment living: Advanced human life support enclosed system final report*. American Astronautical Society: San Diego, CA.

Reed, J. H., Jr., & White, S. C. (1963). *Habitability in space stations*. Retrieved July 1, 2020, from https://ntrs.nasa.gov/search.jsp?R=19630008464

Suedfeld, P., & Mocellin, J. S. (1987). The "sensed presence" in unusual environments Peter Suedfeld. *Environment and Behavior, 19*(1), 33–52.

Suedfeld, P., & Steel, G. D. (2000). The environmental psychology of capsule habitats. *Annual Review of Psychology, 51*, 227–253.

Wise, J. (1988). *The quantitative modelling of human spatial habitability*. Ames Research Center, USA: NASA Contractor Report, 1988. 177501. Retrieved May 15, 2020, from https://core.ac.uk/download/pdf/42829942.pdf

Chapter 2
Habitability: From Place to Space

This chapter summarizes the evolution of the concept of 'Habitability' as it has been envisioned, approached and addressed within the context of extraterrestrial human habitation. It outlines the initial defining characteristics identified from research in terrestrial ICEs (Isolated, Confined and Extreme Environments), traces the evolution of concern with place (location characteristics) to habitation space (built environments) and summarizes the expansion of focus from individual living spaces to those encompassing whole communities that are driven by the demands of the extraterrestrial environment.

2.1 Habitability as PLACE

Today, there is no perfect analogue[1] for extraterrestrial habitats. Although "a *stay at [the Antarctic research station] Concordia has many similarities with a long-duration spaceflight and to future exploration missions [...]*" (ESA 2010), similarities and differences are not the same across different environments. Each isolated and confined environment (extraterrestrial orbital and planetary habitats, polar bases, simulators, etc.) has its own strengths as well as limitations as an analogue environment for the development of future extraterrestrial habitats. But none can be 100% directly compared with the other. Therefore, any investigation into what a habitat should be like for extraterrestrial locations will be, at best, an approximation. For example, even in analogues where conditions limit some critical resource like the availability of a breathable atmosphere (e.g., submarines or underwater habitats), all terrestrial analogues exist in a 1 g field (i.e., 1 gravity or Earth normal gravity). Thus, it will not be possible to *ever* simulate how a microgravity environment affects group

[1]Analogue: "something that is similar to or can be used instead of something else" (Cambridge dictionary 2020).

S. Häuplik-Meusburger, S. Bishop, *Space Habitats and Habitability*, Space and Society, https://doi.org/10.1007/978-3-030-69740-2_2

and individual functioning on Earth. *Does that mean all analogues will be equally incapable of helping us determine how to build effective living spaces for off-earth locales? Or is every analogue so unique that there are no generalizable lessons learned that can be garnered from each?* Despite the aforementioned caveats, the answers to both questions is a resounding 'No'! As Peter Suedfeld, a noted space psychologist observed, "*Analogues were neither completely invalid nor completely useless*" (Suedfeld 1991).

The history of space travel shows that much has been learned from a technological point of view. Life support systems, food systems and the development of new materials are only a few examples of how much has been achieved in the last 50 years. But, when it comes to living in space, knowledge transfer from one environment to the other can become tricky, if not dangerous, for mission success and people's lives. Initially, the focus of early spacecraft (such as Mercury and Apollo programs) was on survivability in which engineering approaches were well equipped to deal with meeting requirements. Provisions of adequate hardware and software systems were primarily a highly specialized technological puzzle. When faced with the needs for the highly variable and, sometimes unpredictable, 'wetware'[2] (i.e., the human participants), the solutions could not be guided by a set of specifications, blueprints or testable processes. To the dismay of much of the engineering and human factors community, social scientists could not provide definitive and generic 'design requirements' to ensure human health and wellbeing. Thus, not only was the solution impossible to achieve, the nature of the 'problem' was similarly unknown. Yet, there undeniably was a problem!

The journey to this realization was a long one. Before we could define and design an effective living envelope[3] for ICEs, we had to understand the greater environment in which that habitat would be located and the demands and requirements of those inhabiting that place. The earliest efforts were directed at simply testing various elements of life support in any available extreme environment under the assumption that the same functionality would be experienced in the same way regardless of location or mission characteristics. The focus was on the demands the environment placed on the constructed structure with almost total disregard to how inhabitants would experience those structures. Just as environmental conditions modified and defined demand characteristics of terrestrial habitats (e.g., submarines require pressurized vessels, polar bases provisions for maintaining heating, equatorial provisions for maintaining cooling), so, too, will the environmental characteristics of extraterrestrial habits similarly define demand characteristics for survival. The more consistent the demand characteristics of place (see Table 2.1) are, logically the easier it should be to address the demand characteristics of the living space. One could argue that the lack of breathable atmosphere, presence of micro or zero-gravity, high radiation exposure, and absence of vegetation and other living forms that

[2]Definition of wetware: the human brain or a human being considered especially with respect to human logical and computational capabilities (Merriam Webster, 2020).

[3]Remember that in most ICE humans can only survive with a protecting envelope.

Table 2.1 Demands and impacts based upon the environment and the actual mission

DEMANDS on habitation system	IMPACTS on habitation system
Demands based upon environmental characteristics ≙ place characteristics	Life support system; restraints for working in microgravity; shelter in case of solar particle events or dust storms; greenhouse to grow food; Etc.
Demands based upon Mission characteristics ≙ expanded place characteristics	Mission goals and tasks; flyby or landing; duration (short or long); Crewsize and composition; Etc.

ubiquitously define all extraterrestrial locations on the near horizon certainly constitute a consistent set of environmental characteristics. Using existing functional operational outpost base designs as templates, multiple simulation facilities have been constructed, situated in various challenging environments and put into service. Yet, habitats in the same location produced different outcomes with different crews. And somewhat different habitats in different environments produced similar results. It became painfully clear that **the variable element was the human inhabitants**. So, the focus shifted to finding the 'right stuff' personality that could effectively adapt to demands of the ICE.

The breakthrough came when researchers trying to identify the 'right stuff' for astronaut crew selection realized that there was *no one right stuff,* but, rather, a **'best fit' personality** for *different mission profiles*. Tagged the 'select-in' criteria (as opposed to the 'select-out' criteria for psychological pathology), research on personality and psychological functioning in crewmembers across various ICEs (Bishop et al. 1998; Bishop and Primeau 2001; Herring 1997; Kahn and Leon 1994; Kanas 1985; Leon 1991; Sandal et al. 1996; Morphew and Maclaren 1997) concluded that it was *mission characteristics* that were, in fact, the drivers for determining 'best fitness'. Therefore, the 'best fit' crewmember could be identified by mission characteristics, e.g., nature of the tasks (daily achievement of specific goals versus station-keeping), risk (mutable or fixed) or duration (short vs. long). These same parameters, by extension, defined the livability demands that must be addressed in the ensuing habitat because these were the characteristics that impacted human perception and activities and, therefore, the *experience* of the habitat. Thus, the same facility and structure could be experienced very differently in a short duration, low risk, and task-driven mission profile compared to a long duration, high risk, in-situ residence mission profile.

In addition to place characteristics, it is the successful match between mission generated adaptation needs and the experience attainable by the living envelope of the space about us that ultimately defines the success of a habitat. It was these emerging factors that helped reshape the approach to designing effective inhabitable space for extraterrestrial environments. Not only do different mission types with particular requirements and constraints result in different mission architecture and spacecraft design (cf. Larson and Pranke 1999), but they also result in different and more powerful habitation design.

2.2 From Exposure to Experience

> One important fact, which has emerged during decades of research, is that in the study of capsule environments there are few main effect variables. Almost every outcome is due to an interaction among a host of physical and social environmental variables and personality factors. Thus, although we conceptually deconstruct the situation into particular sources of variance, we must remember that how people experience an environment is more important than the objective characteristics of the environment. (Suedfeld and Steel 2000, pp. 227–253)

The realization that place characteristics could be reorganized according to mission characteristics rather than environmental characteristics was a significant leap forward. It moved the focus from '*What is special about each of these places?*' to '*What needs do inhabitants in all these places have in common?*' The lessons learned (see Box 2.1) reflected several decades of emergent general findings emphasizing the importance of duration, true confinement and isolation and real risk that have substantially redefined how we approach developing effective living spaces for extraterrestrial habitats.

Box 2.1 Lessons Learned from Analogue Research

- Those exemplified by *short term habitation* durations, e.g., field expeditions such as the Arctic Mars Analogue Svalbard Expeditions (AMASE), annual NASA-funded expeditions that use various field sites on the Svalbard archipelago (Norway), or the two-week missions sponsored by the Mars Society's Mars Desert Research Station in Utah, do not provide for true emergence of societal norms. How groups define the usability of space reflects both function and psychosocial needs of the individual and group. A place for eating is functional but if it also serves as a place for bonding and fellowship, it becomes supportive of greater wellbeing. The former can be predefined; the latter is emergent and needs *time* for development.
- Locations that do not enforce *long-term true confinement or isolation* also lack critical developmental pressures, e.g., space shuttle missions, remote teams on oil rigs or oil cargo transports. Transitional places are temporary spaces and provoke a very different spectrum of human tolerance and adaptation than those that represent long-term inhabitation.
- Analogues in which *real risk* is lacking do not engender the same social dynamics as one in which interdependencies represent real resources for personal and group safety. The criticality of a built environment's characteristics affects human attitudes and values towards that structure. For example, a leak in an underwater habitat is of far more significance than a leak in the roof in Antarctica. The latter may represent a significant threat but it isn't the immediate threat of the former. Perceptions of enclosure are directly proportional to the degree of threat the external environment poses to the habitat.

A defining thread across these critical mission characteristics is the emphasis on *perceived or experienced impact* on inhabitants. It is human perception of risk, confinement, isolation, and duration that defines the utility and critical features of inhabitable space. Analogues where these factors are compromised or lacking pose lesser utility in what that location can teach us about human behavior and the role of effective inhabitable structure in human adaptation. It is important to recognize that a physically superior location (e.g., Antarctica) may be significantly less useful if certain parameters are missing or undermined. For instance, perception of the mutability of consequences has persistently been identified as critical. Rescue from a habitat situated on the campus of a university or space agency is taken as a given whereas the same need from Antarctica during the winter or the International Space Station is fraught with greater uncertainty and risk. And rescue from a lunar base or from Mars will be even greater still. **The greater the perceived risk, the greater will be the sense of confinement and isolation. The greater the perceived isolation, the greater the challenges placed upon the crew and the habitat.** This was confirmed by the reports of the first American astronauts on the Mir space station. Thagard, Blaha and Linenger had severe feelings of isolation and problems with adaption to the physical and, in particular, the social environment (see Sect. 7.4.2).

With the realization that it was the interaction between place characteristics and mission characteristics that shaped the demands upon human inhabitants, the focus shifted from designing for the unique environmental characteristics to the unique needs of *inhabitants* situated in extreme environments by exploring factors that made a difference in how the same habitat could be experienced across different mission parameters and different environments. Experience is shaped by the space that surrounds us, is presented to us, and elicits a response from us. In short, experience is malleable and can be deliberatively directed, for good or ill. What any successful habitat must achieve is the experience of safety, support and sustainment. Given the disparate characteristics of available analogue environments and of those proposed extraterrestrial habitats on the near horizon (L5, Lunar, Mars), Suedfeld (1991) proposed five key principles to improve the efficacy of research in analogue environments to inform us as to how to build for extraterrestrial habitats (see Table 2.2).

This focus on how inhabitants experienced the built environment designated for ICEs was a fundamental shift in perspective. We have to remember that the approach to early spaceflight was purely functional. There were, for example, no reports of

Table 2.2 Key principles to improve the efficacy of research in analogue environments (Suedfeld 1991)

Principles
Principle 1: Researchers should think in terms of experiences within environments rather than of environmental characteristics;
Principle 2: Researchers should study differences and similarities between experiences, which are not the same as those between environments;
Principle 3: Analogies should be based on similarities of experience, not necessarily of environment;
Principle 4: Research should look at systematic links between personality factors and experience; and
Principle 5: Experience is continuous and integrated.

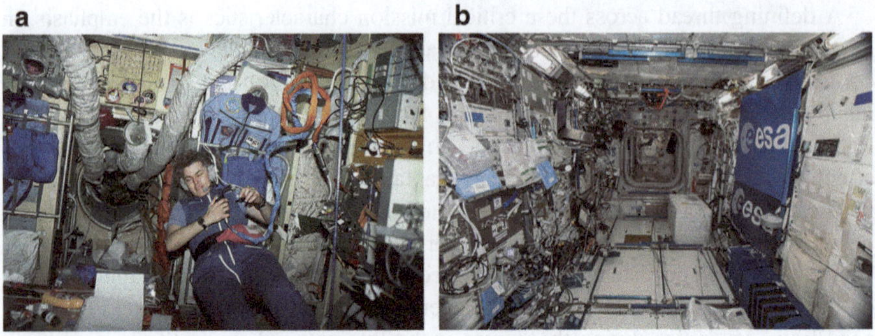

Fig. 2.1 (**a**) Shannon Lucid aboard Mir space station. Launched aboard the STS-76, the third Shuttle/Mir docking mission, in March 1996, to join the Mir crew in the orbiting laboratory. Astronaut Lucid returned to Earth aboard STS-79 in September 1996. At the time, Lucid set the U.S. record for the longest stay in space of 188 days (Credit: NASA). (**b**) Interior of the International Space Station. A view of the European Space Agency Columbus Lab Module, looking across into the Japanese Experiment Module, 2018 (Credit: NASA)

space sickness from Mercury and Gemini astronauts. But with Apollo and Skylab, it went up to 50% (Crampton 1990). In the case of cosmonaut Valentina Tereshkova,[4] her report of motion sickness was directly attributed to her being a woman as women had been seen as more vulnerable, thus being more likely to experience space motion sickness and men more likely to experience re-entry sickness upon return to Earth. Heretofore, a largely functional engineering approach dominated in which the essential features were providing for survivability (air, water, power, waste management, food). Hence, the interiors were, and often still are, characterized by congested, highly complex environments exposing mechanical, electrical and logistical hardware to provide easy access for repairs with no 'wasted' resources dedicated to psychological functioning (see Fig. 2.1a, b).

Terrestrial analogues were no better. Early evaluations for astronaut selection drew upon a history of sensory deprivation research initially begun by the military throughout the 1950s and 1960s to address performance concerns about two-person crews confined to armored vehicles for "long durations" and continued most notably through the series of studies conducted by Zubek (1969). Initially, it was believed that space would represent a significant loss of normal sensory stimulation due to isolation from people, reduction in physical stimulation, and restricted mobility. Thus, sensory deprivation chambers were argued to be good analogues for astronauts (Flaherty 1961). Selection procedures, therefore, included stints in dark, small, enclosed spaces for several hours to observe how potential astronauts handled the confinement and loss of perceptual cues. As American Astronaut, Dr. Bernard Harris, the first African American to walk in space, recounts, "*They put me in this little box where I couldn't move or see or hear anything. As I recall, I fell asleep after*

[4] *"Валентина Терешкова: чьей воле покорялась женщина, покорившая космос" [Valentina Tereshkova: the Woman who Conquered Space]. RIA Novosti (in Russian). 16 June 2006. Retrieved 3 April 2016.*

a while until the test ended" (Harris 1995–96).[5] However, even as successful orbital flights and subsequent station experience disproved the sensory deprivation model, the employment of utilitarian, austere functional built environments continued as the templates for subsequent ICE human habitats both on- and off-earth.

2.2.1 The Early Years: Moving from Pure Exposure to Experience Through Chamber Research

The first systematic attempts to investigate psychological adaptation to isolation and confinement in simulated confined environments, were conducted in sensory deprivation chambers beginning in the 1950's and continuing throughout the 1960s and early 1970s. Chamber research, as it was to become known, eventually encompassed a variety of artificial, constructed environments whose raison d'être was control over all factors not specifically under study. For instance, volunteers were frequently simply confined in closed rooms for several days, subjecting them to sleep deprivation and/or various levels of task demands by having them complete repetitive research tasks to evaluate various aspects of performance decrements (Haythorn and Altman 1966; Altman 1973).

Early ground-based chamber studies were conducted by NASA to prepare for upcoming missions and test systems, equipment and procedures. Table 2.3 lists a number of early-ground based simulations that already included some issues related to the actual habitat. It was during one of those simulations the 1967 accidental deaths of the Apollo 1 crew occurred because the cabin was pressurized with pure oxygen and the door could not be opened from the outside (Box 2.2). The atmosphere was changed for the second test to 60% oxygen and 40% nitrogen at 15 psi (Lange et al. 2002). Those simulations were conducted in an environment as close to the actual mission parameters as possible and the astronauts wore biomedical vests. The test of the 'acceptability' of the habitat lead to a number of improvements, such as the re-design of the Apollo space suits and the urine system of Skylab, because it was discovered to be too small.

Box 2.2 Apollo 1: The Mission that Never Flew

The first manned mission towards the Moon—an Earth-orbiting mission—was planned to launch on February 21, 1967. A few weeks before, the whole crew was killed by a cabin fire during a launch rehearsal test at Cape Kennedy. The mission was initially designated AS-204 but was renamed to Apollo 1 in honor of the crew: Virgil (Gus) Grissom (Command Pilot), Ed White (Senior Pilot),

(continued)

[5] B. Harris, personal communication, thesis committee member (1995–1996).

Box 2.2 (continued)
and Roger B. Chaffee (Pilot). It would take another 18 months, and extensive redesigns, before another astronaut would enter a rocket. Eventually, the first Apollo mission to get to space was Apollo 7 on October, 11 in 1968.

Further Reading: https://www.nasa.gov/mission_pages/apollo/apollo-1.

Table 2.3 Examples of early ground-based chamber studies (Lange et al. 2002; Wright and Jaques 2002)

Early ground-based simulations	Objective	Crew	Architecture	Facility	Relevant habitability issues
MOLAB (1965), General Motors for NASA	Extended lunar mission manned vehicle	2 males	Pressurized rover	US geological survey	Proved astronauts could stay in it for 18 days; Acceptability of cabin and suit
Apollo Ground Based Tests (1968)	Manned vacuum chamber tests to prepare to fly to the moon	3 males	CSM with 310 ft^3 (8.8 m^3) of pressurized volume	Johnson Space Center, on two Apollo command/service modules (CSMs)	
Regenerative Life Support Study by NASA Langley Research Center (1970)	60 and 90-day manned test of regenerative life support system	4 males	Chamber was 12 ft. (3.6 m) in diameter and 40 ft. (12.2 m) long, with 4.100 ft^3 (116 m^3), a 160 ft^3 (4.5 m^3) airlock, and 2 smaller airlocks which were 18 in (0.46 m) in diameter, 10 psi	McDonnell Douglas astronautics company, Huntington Beach, CA	The acceptability of the habitat, crew training evaluations, and computer assistance scheduling were evaluated
The Skylab medical experiments altitude test (SMEAT) (1970s)	56-day simulation of a Skylab mission with "full-up dress rehearsal"	3 males	Two-floor habitat with 20 ft. diameter with similar systems as Skylab (collapsible shower, water and waste management)	Johnson Space Center, NASA	Tests of the urine system demonstrated specific problems, such as being too small, resulting in redesign. Life sciences included lower body negative pressure, oral health, habitability/crew quarters, and crew training

Later, specially constructed confinement laboratories (see Chap. 4), housed small groups of three to six individuals in programmed environments for weeks to months of continuous residence to address a variety of space/science related human bio-behavioral issues concerning group dynamics (e.g., cohesion, motivation, effects of joining and leaving established groups), performance and work productivity, communication patterns, team cooperation, and social habitability factors. Other types of chambers have been utilized to evaluate various countermeasures, especially for stress and anxiety provoked by physiological stressors expected to be present in the space environment. For instance, in addition to their Lunar Palace simulation habitat, Chinese behavioral scientists were the latest to employ the use of centrifuge training to test the efficacy of guided imagery to reduce anxiety, heart rates and heart rate variability (Jing et al. 2011). Critics of this chamber facility approach point out that such environments constitute *confinement* but not necessarily isolation. The stress of isolation is far more complex than that of confinement and involves a psychological dimension of separateness that may not be inherent in artificial chamber studies. "*Isolation [...] refers to situations in which even that level of social contact and support is lost in highly restrictive settings where inmates have little or no contact with others*" (Clayton 2012, p. 320).

The epitome example of chamber research may be the series of five hyperbaric-chamber studies, dubbed the MARS 500 (see also Sect. 4.3.2), a cooperative project between ESA and the Russian Institute for Biomedical Problems (IBMP), utilizing their facilities in Russia, and investigating psychosocial functioning, in which groups were confined for periods lasting from 28 to 520 days (Sandal et al. 1995; Sandal 2004). Full mission protocols specifying all medical, technical, and operational parameters approximating expected living conditions of astronauts on a space station (and surface habitat) were used. The studies were intended to evaluate the efficacy of various psychosocial monitoring and assessment techniques for implementation on real space missions, as well as to investigate persistent occurrences of communication and interaction breakdowns between on-orbit teams and Mission Control anecdotally reported from space (Kanas et al. 2000; Sandal et al. 1995; Gushin et al. 1996).

Such simulation missions have also provided opportunities to learn from mission failures. During an earlier long simulation mission also conducted at IBMP in 1999 (SFINCSS'99) (Sandal 2004), then the longest mission to date, a crew meltdown occurred. Too many New Year's Eve vodka shots precipitated an incident involving unwanted sexual advances between one of the resident long duration male crew members and an international female team member on the shorter duration 'visiting' crew. This provoked a fistfight between male members. The fracas provoked reactions internally and across involved agencies as well as publicly that resulted in the two crews being physically isolated from each other for a period and one international team member leaving the project mid-mission. A mediation team was brought in to reconcile the teams and although the remaining crew members stayed and finished the mission, it was undeniable that the escape 'solution' or face-to-face mediation approach would not have been available in a real extraterrestrial mission (see Box 2.3). Subsequent mission design took special note of the contributing factors involving different cultural attitudes, agency policies, crew norms and operational stresses, and sought to mitigate these. The Mars 500 simulations in particular were focused on evaluating how successful those mitigation strategies were.

Box 2.3 When Simulation Study Conditions Become a Threat…

The IBMP SFINCSS mission reported here was one we studied in my Lunar Base study. We were shocked to learn that the female visiting participant in the IBMP had slept with a knife she'd stolen from the mess in order to protect herself at night!

My best recollection of the reported circumstances is that this was a simulation mission where some new crew members joined up half way through, and the female participant was in that new group. The guy who hit very hard on her was a Russian fellow who was a member of the long, full term mission crew. This obvious problem was glossed over as a (1) Cross-Cultural misunderstanding, and (2) the kind of a conflict where new crew members are suddenly introduced to an established group, disrupting a situation where interpersonal dynamics are already sorted out.

It didn't come out until the post mission debriefings that she had stolen the kitchen knife and slept with it, indicating how threatened she felt.[6]

I see so many parallels in this case to reports we read all the time about sexual harassment of women military members, and it's a pretty strong indicator that 'military organization and discipline' is not good enough to stop it from happening. Lesson: We're not going to avoid such things by either (a) individual crew member selections, especially with cross-cultural crews, or (2) strict established rank and reporting hierarchies.

Personal communications, Jim Wise, 2021.

A number of opportunities and advances came from these studies, e.g., evaluating the efficacy of communication training for space teams or the opportunity to examine factors involved in an unplanned meltdown between crews precipitated by differences in cultural attitudes and norms about genders, authority, and control (Sandal 2004). However, skepticism regarding the verisimilitude of studies in which discontented members can simply quit has continued to raise real concerns as to how generalizable the findings from chamber studies are to extraterrestrial missions.

2.2.2 *Seeking Real Risk, Isolation and Confinement Through Operational Habitats in Isolated and/or Extreme Environments*

From chamber studies situated in agency or university campuses, the next extension was to study teams in real operational environments. The emphasis was largely on

[6]https://www.theglobeandmail.com/news/nationa/canadians-harrassment-complaint-scorned/article25458615/.

investigating small group dynamics and human performance in habitats like the Antarctic bases (e.g., McMurdo, Troll Station, Concordia, South Pole) (see also Chap. 5) as well as expeditions conducted in various challenging environments, e.g., treks to the poles, climbing mountains, crossing deserts.

Inarguably, Antarctica provides the best terrestrial analogue for extraterrestrial habitation studies (see also interview with Alexander Kumar in Box 3.6 and 5.2). Not only is the environment inimical to human survival, the duration of environmentally enforced isolation and confinement (~9 months) is the most extreme on Earth. Because of this fact, while there are other polar bases in the Arctic and subarctic, the bulk of sustained long duration psychological research has been conducted in Antarctica (Lilburne 2005; Sandal et al. 2018). While the exact number of permanent stations (those occupied year-round) throughout the Antarctic and sub-Antarctic regions varies depending on national funding, base conditions, and political situations, the count has fluctuated between 30 and 70, operated by the 48 signatory nations to the Antarctic Treaty, with populations running from 4 to 1100 men and women in the summer to 4 to 250 during the winter months.[7] The base populations vary from mixed-gendered crews to male-only crews, from intact families (Chile) to unattached singletons, for assignments that last from a few months to 3 years.

The first systematic psychological study of 85 men wintering over in Antarctica was conducted in 1958 by C. S. Mullin, H. Connery, and F. Wouters, after the 1956–57 International Geophysical Year (IGY) produced the first permanent bases in Antarctica (Mullin et al. 1958). As with the majority of psychological and medical research in various polar habitats through the 1980s, the focus was on the physiological changes evidenced in winter-over adaptation. Mullin et al.'s (1958) study was the first of many to identify the Antarctic fugue state later dubbed the "big-eye,"[8] and most recently referred to as psychological hibernation (Sandal et al. 2018), characterized by pronounced absent-mindedness, wandering of attention, and deterioration in situational awareness that surfaced after only a few months in isolation. Similarly, later studies addressing psychosocial factors tended to focus on the negative or pathological problems of psychological adjustment to Antarctic isolation and confinement, with persistent findings of depression, hostility, sleep disturbance, and impaired cognition, which quickly came to be classified as the "winter-over syndrome" (Gunderson 1973, 1974; Strange and Klein 1974). However, it has not all been negative. Sprinkled among Antarctic research have been findings that also report positive, and even salutogenic aspects of the winter-over experience in which winter-overers have reported enhanced self-growth, positive impacts to careers, and opportunities for reflection and self-improvement (Mullin

[7]For more information on Polar Bases go to: (https://www.coolantarctica.com/Community/antarctic_bases.php).

[8]Another term for insomnia due to the lengths of the days and nights in the polar regions (cf. Stephenson 2009, p. 113).

1960; Taylor and Shurley 1971; Wilson 1965; Palinkas 1986; Oliver 1991; Nielsen and Vollers 2001; Suedfeld 2005).

The emergence of both positive and negative effects on inhabitants of ICEs was illuminated by one of Antarctica's most prolific researchers, Dr. Larry Palinkas, in his analysis of 1100 Americans who wintered over between 1963 and 2003, representing over four decades of research in Antarctica. From this extensive meta-analysis, he proposed four distinct characteristics of psychosocial adaptation to isolation, confinement, and the extreme environment experienced in Antarctica (Box 2.4).

Box 2.4 Characteristics of Psychosocial Adaption to Isolation, Confinement, and the Extreme Environment (Palinkas 2003)

1. Adaptation follows a seasonal or cyclical pattern that seems to be associated with the altered diurnal cycle and psychological segmentation of the mission.
2. Adaptation is highly situational. Because of unique features of the station's social and physical environment and the lack of resources typically used to cope, baseline psychological measures are not as good predictors of depressed mood and performance evaluations as are concurrent psychological measures.
3. Adaptation is social. The structure of the group directly impacts individual wellbeing. Crews with clique structures report significantly more depression, anxiety, anger, fatigue, and confusion than crews with core-periphery structures.
4. Adaptation can also be "salutogenic," i.e., having a positive effect for individuals seeking challenging experiences in extreme environments.

In his analysis, Palinkas found evidence that the winter-over experience was associated with reduced subsequent rates of hospital admissions when participants returned home. He and others have speculated that the experience of adapting to the isolation and confinement, in general, improved an individual's self-efficacy and self-reliance and engendered coping skills that they used in other areas of life to buffer subsequent stress and resultant illnesses (Palinkas 2003; Suedfeld 2005).

In 2006, Sandal et al. conducted an extensive review of the literature on psychosocial adaptation by a variety of groups in extreme environments including polar work groups, expedition teams, Antarctic bases, simulation, and space crews (Sandal et al. 2006). The comparison across both simulation and operational mission profiles and polar groups covering a wide spectrum of mission structures and goals dramatically highlighted critical differences regarding crucial mission characteristics, especially those involving tasks and goals as well as short and long duration that clearly identified habitation characteristics as *different* from expedition characteristics.

Applying the emerging framework of select-in personality characteristics for crew members, findings reinforced the realization that there was a distinct difference in individuals who were 'best fit' for expeditionary needs (i.e., short duration, high task focused, clearly defined milestones, specific team roles, and recognizable performance metrics) compared to those that were 'best fit' for missions in which long duration habitation demands (e.g., high interpersonal skills, autonomy, self-reliance, flexibility, tolerance for ambiguity, monotony and boredom) would be in play. That insight, subsequently, propelled the momentum to explicitly look at the **habitation fit** to the **experience driven needs** of those living in ICEs, completing the journey from the focus on pure exposure to experience.

References

Altman, I. (1973). An ecological approach to the functioning of isolated and confined groups. In J. Rasmussen (Ed.), *Man in isolation and confinement* (pp. 241–270). Chicago: Aldine.

Bishop, S., Hilliard, P., D'aunno, D., Harris, B, Jones, J., Rossum, A. and Stanford. (1998) *Biobehavioral Life Support Issues for Living and Working in Space*, The Third International Conference on Life Support and Biosphere Science, January 11–15, Lake Buena Vista, Florida.

Bishop, S. L., & Primeau, L. (2001). Assessment of group dynamics, psychological and physiological parameters during polar winter-over. In *Proceedings for the Human Systems Conference*, Nassau Bay, Texas, June 20–22 2001.

Clayton, S. D. (2012). *The Oxford handbook of environmental and conservation psychology*. New York: Oxford Press.

Cool Antarctica. *Antarctic stations – Bases – Currently occupied*. Retrieved from https://www.coolantarctica.com/Community/antarctic_bases.php

Crampton, G. H. (1990). *Motion and space sickness*. Boca Raton, FL: CRC Press.

European Space Agency (ESA). (2010). *Concordia calling*. Retrieved 1 July 2020, from https://phys.org/news/2010-01-concordia.html

Flaherty, B. E. (Ed.). (1961). *Psychophysiological aspects of space flight*. New York: Columbia University Press.

Gunderson, E. K. E. (1973). Individual behavior in confined or isolated groups. In J. Rasmussen (Ed.), *Man in isolation and confinement* (pp. 145–166). Chicago: Aldine.

Gunderson, E. K. E. (1974). Psychological studies in Antarctica. In E. K. E. Gunderson (Ed.), *Human adaptability to Antarctic conditions* (pp. 115–131). Washington, DC: American Geophysical Union.

Gushin, V. I., Kolintchenko, V. A., Efimov, V. A., & Davies, C. (1996). Psychological evaluation and support during EXEMSI. In S. Bonting (Ed.), *Advances in space biology and medicine* (pp. 283–295). London: JAI Press.

Haythorn, W., & Altman, I. (1966). Personality factors in isolated environments. In M. Trumbull (Ed.), *Psychological stress: Issues in research* (pp. 363–386). New York: Appleton-Century-Crofts.

Herring, L. (1997). Astronaut draws attention to psychology, communication. *The Journal of Human Performance in Extreme Environments, 2*(1), 42–47.

Jing, X., Liu, F., Wu, B., Miao, D., et al. (2011). Guided imagery, anxiety, heart rate, and heart rate variability during centrifuge training. *Aviation, Space, and Environmental Medicine, 82*(2), 92–96.

Kahn, P. M., & Leon, G. R. (1994). Group climate and individual functioning in an all-women Antarctic expedition team. *Environment and Behavior, 2*(5), 669–697.

Kanas, N. (1985). Psychosocial factors affecting simulated and actual space missions. *Aviation, Space and Environmental Medicine, 56*, 806–811.

Kanas, N., Salnitskiy, V., Grund, E. M., et al. (2000). Social and cultural issues during shuttle/Mir space missions. *Acta Astronautica, 47*(2–9), 647–655.

Lange, H. W., Sauer, R. L., Feeback, D. L., & American Astronautical society (AAS) (Eds.). (2002). *Isolation. NASA experiments in closed-environment living. Advanced human life support enclosed system*. San Diego, CA: Univelt Inc.

Larson, W. J., & Pranke, L. K. (1999). *Human spaceflight: Mission analysis and design (space technology series)*. New York: McGraw-Hill.

Leon, G. R. (1991). Individual and group process characteristics of polar expedition teams. *Environment and Behavior, 23*(6), 723–748.

Lilburne, L. (2005). *Shrinks on ice: A review of psychosocial research in Antarctica*. Graduate certificate in Antarctic studies. Retrieved 1 July 2020, from https://ir.canterbury.ac.nz/bitstream/handle/10092/13978/Lilburne_L_Lit.Review.pdf?sequence=1 accessed 5/22/2020

Morphew, M. E., & Maclaren, S. (1997). Blaha suggests need for future research on the effects of isolation and confinement. *Journal of Human Performance in Extreme Environments, 2*(1), 52–53.

Mullin, C. S. (1960). Some psychological aspects of isolated Antarctic living. *American Journal of Psychiatry, 117*, 323–325.

Mullin, C. S., Connery, H., & Wouters, F. (1958). *A psychological-psychiatric study of an IGY Station in Antarctica*. Report prepared for the U.S. Navy, Bureau of Medicine and Surgery, Neuropsychiatric Division.

Nielsen, J., & Vollers, M. (2001). *Ice bound: A Doctor's incredible battle for survival at the south pole*. New York: Hyperion.

Oliver, D. (1991). Psychological effects of isolation and confinement of a Winter-over Group at McMurdo Station, Antarctica. In A. A. Harrison, Y. A. Clearwater, & C. P. McKay (Eds.), *From Antarctica to outer space: Life in isolation and confinement* (pp. 217–228). New York: Springer.

Palinkas, L. A. (1986). Health and performance of Antarctic winter-over personnel: A follow-up study. *Aviation, Space, and Environmental Medicine, 57*(10, sect 1), 954–959.

Palinkas, L. A. (2003). *On the ICE: Individual and Group Adaptation in Antarctica*. Retrieved June 12, 2007, from http://www.sscnet.ucla.edu/anthro/bec/papers/Palinkas_On_The_Ice.pdf

Sandal, G. M. (2004). Culture and crew tension during an international Space Station simulation: Results from SFINCSS'99. *Aviation, Space, and Environmental Medicine, 75*(7, section 2, Suppl), C44–C51.

Sandal, G. M., Vaernes, R., Ursin, H., et al. (1995). Interpersonal relations during simulated space missions. *Aviation, Space, and Environmental Medicine, 66*(7), 617–624.

Sandal, G. M., Vaernes, R., Bergan, P. T., Warncke, M., Ursin, H., et al. (1996). Psychological reactions during polar expeditions and isolation in hyperbaric chambers. *Aviation, Space, and Environmental Medicine, 67*(3), 227–234.

Sandal, G. M., Leon, G. R., & Palinkas, L. (2006). Human challenges in polar and space environments. *Review Environmental Science and Biotechnology, 5*, 281–296. https://doi.org/10.1007/s11157-006-9000-8.

Sandal, G. M., van deVijver, F. J. R., & Smith, N. (2018). Psychological hibernation in Antarctica. *Frontiers in Psychology*. https://doi.org/10.3389/fpsyg.2018.02235

Stephenson, J. (2009). *Crevasse roulette – The first trans-Antarctic crossing 1957–58*. Kenthurst: Rosenberg, National Library of Australia Cataloguing-in-Publication.

Strange, R., & Klein, W. (1974). Emotional and social adjustment of recent U.S. winter-over parties in isolated Antarctic Station. In O. G. Edholm & E. K. E. Gunderson (Eds.), *Polar human biology: The Proceedings of the SCAR/IUPS/IUBS Symposium on Human Biology and Medicine in the Antarctic* (p. 410). Chicago: Year Book Medical Publications.

Suedfeld, P. (1991). Groups in isolation and confinement: Environments and experiences. In A. A. Harrison, Y. A. Clearwater, & C. P. McKay (Eds.), *From Antarctica to Outer Space: Life in isolation and confinement* (pp. 135–146). New York: Springer.

Suedfeld, P. (2005). Invulnerability, coping, salutogenesis, integration: Four phases of space psychology. *Aviation, Space, and Environmental Medicine, 76*(6 Suppl), B61–B73.

Suedfeld, P., & Steel, G. D. (2000). The environmental psychology of capsule habitats. *Annual Review of Psychology, 51*, 227–253.

Taylor, A. J. W., & Shurley, J. T. (1971). Some Antarctic troglodytes. *International Review of Applied Psychology, 20*(2), 143–148.

Wilson, O. (1965). Human adaptation to life in Antarctica. In J. Van Meigheim, P. van Oue, & J. Schell (Eds.), *Biogeography and ecology in Antarctica, Monographiae Biologicae* (Vol. 15, pp. 690–672). The Hague: W. Junk.

Wright, M., & Jaques, B. (2002). *A brief history of the lunar roving vehicle*. Retrieved 2020, from https://www.hq.nasa.gov/alsj/MSFC-LRV.pdf

Zubek, J. P. (1969). *Sensory deprivation: Fifteen years of research*. New York: Appleton-Century-Crofts.

Chapter 3
Habitability as SPACE

This chapter describes an **Integrated Habitability Model** of psychological, physiological, sociocultural and spatial habitability considerations. The meaning of the term 'habitability' is fraught with differences in usage, focus, application and definition. Historically, it has evolved from 'habitability as location'[1] to 'habitability as living space'. In application, it is used by different professions with a variety of underlying foci. Thus, a multiplicity of concepts exists. As such, it provides challenges to systematic study as well as constructive dialogue. This chapter gives a systematic overview of the common features and differences of those concepts and argues for a robust inclusive model centered around inhabitant experience of space as a central core.

3.1 A Paradigm Shift Within Habitability Analogue Research

In Chap. 2, the evolution of habitability from a primary focus on provision for survivability in an extreme *place* to provision for *living spaces* that address human experience defined by both place and situation was outlined. The second major paradigm shift in the evolution of habitability as place to living space started when the architectural community began working closely with the human factors community in identifying the design characteristics of proposed long duration extraterrestrial habitats.

Regardless of whether it was for the moon or Mars, in orbit or a long duration vehicle in transit, unique non-terrestrial environmental characteristics (e.g., microgravity) placed an immense challenge on habitation design. Although the immediate unavailability of a few resources (e.g., power, food, water) has been simulated in some terrestrial analogues, many other relevant design characteristics are uncommon

[1] In astronomy and astrobiology the 'habitable zone' is a range of orbits around a star (e.g., our sun) within which a planetary surface can support liquid water given sufficient pressure.

S. Häuplik-Meusburger, S. Bishop, *Space Habitats and Habitability*, Space and Society, https://doi.org/10.1007/978-3-030-69740-2_3

Table 3.1 Relevant design characteristics for space architecture

Typical environmental design characteristics in space architecture	
Not available on Earth	Uncommon
Microgravity conditions high radiation exposure Environmental conditions requiring 100% life support Extreme temperature ranges within hours Threats of solar flares Vacuum Lack of breathable atmosphere Unmodifiable communication time lag Absence of all other lifeforms Complete separation from humanity Earth out of sight	Reliance on a 100% closed loop environmental system (100% recyclable air, water and waste) Limited in-situ-available resources (e.g., power, food, water) Strictly enforced isolation and confinement Enforced acquaintances with a small group of people (microsociety) Inaccessibility of rescue/aid Inaccessibility of communication with outside Requirement for breathable atmosphere (e.g., under-water) Extreme temperatures

Table 3.2 Environmental issues affecting individual and group functioning

Both individual and group functioning is impacted by
– A reliance on technology for life support and performance; – Physical and social isolation and confinement; – High risk and associated cost of failure; – High physical/physiological, psychological, psychosocial, and cognitive demands; – Closed and dependent human-human, human-technology, and human-environment interfaces; – Requirements for team coordination, cooperation, and communication

or non-existent in terrestrial habitats (see Table 3.1). While the consistency of place demands (see also Table 2.1) and common mission characteristics would usually make designing effective habitats easier, having to incorporate demand characteristics that cannot be tested at all or otherwise with high difficulties provides immense challenges.

Layered upon the difficulties in testing physical parameters, comes the psychological and sociocultural habitability considerations for extraterrestrial habitats which are similarly embedded within the complex matrix of the technological and architectural systems. As outlined in Chap. 2, habitability related issues have impacted significantly on human behavior and performance in most challenging environments, especially those characterized by isolation and confinement (Herring 1997; Manzey and Lorenz 1997; Manzey et al. 1995; Morphew and Maclaren 1997). Table 3.2 summarizes the main environmental issues affecting individual and group functioning found across all ICEs (Bishop 2006).

In response to the extraordinary dependency on technology that is inherent in the construction of vehicles and habitats for extraterrestrial environments, a system was needed to categorize levels of technological capability for proposed studies and projects. In order to quantify and qualify risk, as well as capability, NASA and other space agencies established the system of Technological Readiness Levels (TRLs) in the 1970s for the testing of technological systems and elements (see Table 3.3). Since then a number of other methods or standards for the evaluation of the 'maturity' of systems have been developed. In order to address habitability

Table 3.3 Technology readiness level definitions

TRL	Definition
1	Basic principles observed and reported
2	Technology concept and/or application formulated
3	Analytical and experimental critical function and/or characteristic proof-of concept
4	Component and/or breadboard validation in laboratory environment
5	Component and/or breadboard validation in relevant environment
6	System/subsystem model or prototype demonstration in a relevant environment (ground or space)
7	System prototype demonstration in a space environment
8	Actual system completed and "flight qualified" through test and demonstration (ground or space)
9	Actual system "flight proven" through successful mission operations

Table 3.4 Habitation readiness levels and its relation to technology readiness levels (Connolly et al. 2006)

Habitation systems research	Research and design levels	Habitat subsystem technologies should have the following TRL
Habitation systems research (Level 1)	Level 1: Human factors, crew systems, and life support research related to habitation systems	Any TRL
Conceptual and functional feasibility of the technology (Level 1–4)	Level 2: Habitation design and concepts, functional and task analysis	Any TRL
	Level 3: Internal configuration, functional definition and allocation, use of reduced scale models	TRL 6 or higher
	Level 4: Full-scale, low-fidelity mockup evaluations	
Demonstration of the technology (Level 5–6)	Level 5: Full-scale, high-fidelity mockups, human testing and occupancy evaluations	TRL 6 or higher
	Level 6: Habitat and deployment field testing	TRL 7 or higher
Testing of the technology and technology operations (Level 7–8)	Level 7: Pressurized habitat prototype testing	TRL 8 or higher
	Level 8: Actual systems completed and "flight qualified" through test and demonstration	
	Level 9: Actual system "flight proven" through successful mission operations	

requirements and design aspects in correlation to the TRLs, a group of NASA engineers[2] developed the 'Habitation Readiness Levels (HRL)' (see Table 3.4) (Häuplik-Meusburger and Bannova 2016; Cohen 2012).

[2] Jan Connolly, Kathy Daues, Robert Howard, and Larry Toups.

To date, predominantly only partial habitable living and working environments and subsystems in relevant environments have been tested on Earth. Full habitable living and working simulations, such as the Biosphere 2 missions (see Sect. 4.3.1), have been very rare. Most of the existing analogues address the TRLs 1–3 (up to the proof of concept without relation to an environment) and the TRLs 4–6 with a test in a 'relevant' or laboratory environment (e.g., simulated or operational ICEs). TRLs 7–9 are related to the testing (of prototypes) in the real extraterrestrial environment. There has yet to be a full-scale habitability mockup in a space environment at such a high TRL. The closest we have come is the Bigelow Expandable Module (BEAM), which is a prototype for a future deployable space habitat. It was attached as temporary experimental module to the International Space Station (ISS) in 2016 (see Fig. 3.1). The main purpose was to test its durability, but due to its engineering and performance assessment, it was decided to keep it in place until 2028. However, it is being used as storage module.

Virtual simulators are a relatively new category. They have been mainly used by space agencies for astronaut training to prepare them for spacewalking (see Chap. 4). Thus, habitability studies at the highest TRLs have been studies of disparate components on various space stations and polar stations, both of which were constructed with purely functional, utilitarian survivability foci as discussed in

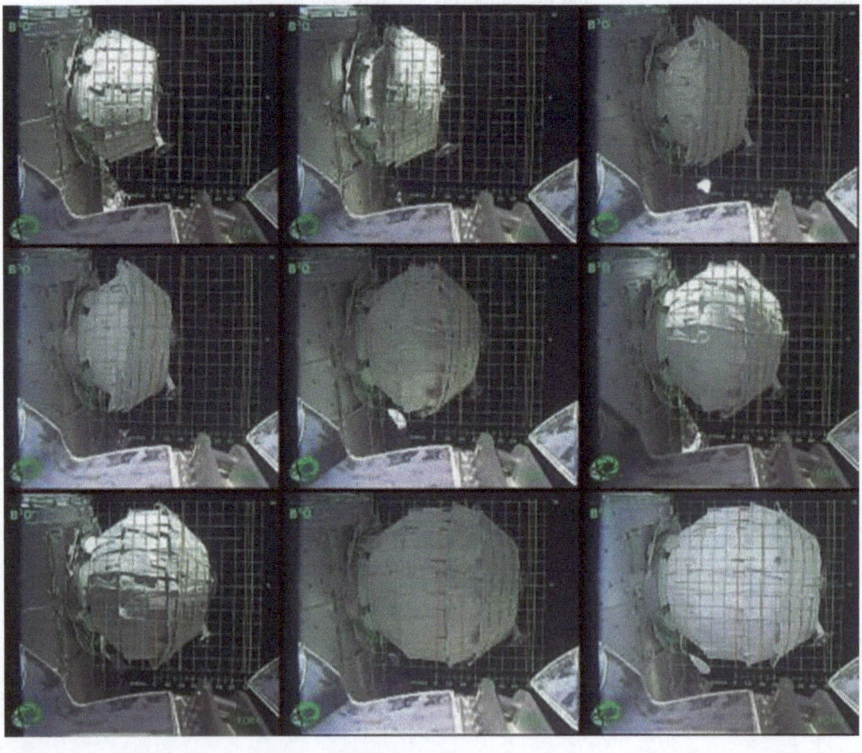

Fig. 3.1 Progression of expansion of the BEAM. After a seven-hour work session, NASA and Bigelow Aerospace were able to get BEAM expanded to full size on May 28, 2016 (credit: NASA TV)

Chap. 2. And therein lies the crux of the problem. Many habitability considerations are critical components for human missions but cannot be tested at the highest technological readiness level on Earth.

Without an actual extraterrestrial testbed, space habitat designers are facing the challenge of **(1) how to identify relevant habitability issues, (2) how to address negative effects with habitability countermeasures and eventually (3) how to evaluate and prove their effectiveness before building and living in the actual habitat.**

3.2 Identifying the 'Relevant' Habitability Issues for Extraterrestrial Environments

A wealth of evidence exists underscoring the contribution of **architecture and interior design as key factors in human health and wellbeing**. Although habitability gained importance in the engineering world, its integration has always led to big arguments between the engineers and designers (Box 3.1).

Box 3.1 Habitability Integration Design Arguments for the Skylab Space Station

In 1968, Raymond Loewy, a world-renowned industrial designer was hired as a habitability consultant for the Saturn-Apollo and Skylab projects. George E. Mueller, NASA Associate Administrator for Manned Space Flight at NASA Headquarters, seems to have been the driving force behind the inclusion of human factors in the design of the Skylab Space Station. Loewy suggested a number of improvements to the layout, such as the implementation of a wardroom, where the crew could eat and work together, the wardroom window, the dining table and the color design, amongst others. The design process was difficult. The engineers regarded crew quarter design as part of their responsibility and the test pilot astronauts who were reviewing the crew quarter concepts adopted the attitude that they cared even less about the design. It took some time until the improvements Loewy suggested were implemented (Compton and Benson 1983).

One controversial subject was the provision of a window in the Skylab Saturn Workshop's Wardroom area. The engineers argued it would weaken the structure, be too expensive, and take too long to develop. At first the window seemed not to be essential to mission success. Mueller asked Loewy for his opinion, on which he replied: "*Not to have a window is unthinkable*!" (Compton and Benson 1983, p. 137). Mueller answered: "*Put in the window*" (Compton and Benson 1983, p. 137) and the window, the wardroom and other changes were authorized.

George Mueller later said of the early mockup: "*Nobody could have lived in that thing for more than two months*" (Compton and Benson 1983, p. 133).

[As published in the book Architecture for Astronauts (Häuplik-Meusburger 2011, p. 2)]

The implementation of the Skylab wardroom window (Fig. 3.2a) is a leading example of the importance of integrating habitability into the design. The window was appreciated by the astronauts, although the astronauts could not always see the Earth as 'proposed'. In many images from the Skylab missions the wardroom window is actually closed (Fig. 3.2b). For astronaut Weitz (Weitz 2000), "the best windows" were in the MDA, because it had windows at a 90-degree angle and thus always provided an Earth view (cf. Häuplik-Meusburger 2011, p. 279). Although not all implementations were as useful as intended (Table 3.5), the Skylab station was probably the first in-space testbed for many activities to be performed in space.

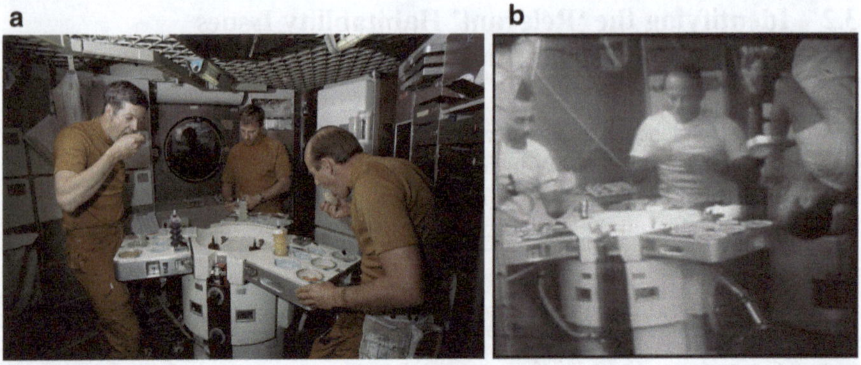

Fig. 3.2 (**a**) Skylab 2 astronauts eat space food in wardroom of Skylab trainer. The three members of the prime crew of the first manned Skylab mission dine on specially prepared Skylab space food in the wardroom of the crew quarters of the Skylab Orbital Workshop (OWS) trainer during Skylab training at the Johnson Space Center. They are, left to right, Scientist-Astronaut Joseph P. Kerwin, science pilot; Astronaut Paul J. Weitz, pilot; and Astronaut Charles Conrad Jr., commander. (credit NASA). (**b**) Skylab 3 astronauts are shown eating in the Orbital Workshop (OWS) wardroom of the Skylab space station in Earth orbit. Astronaut Alan L. Bean (right), commander, illustrates eating under zero-gravity conditions upside down. The two other crewmen are scientist-astronaut Owen K. Garriott (left), science pilot; and astronaut Jack R. Lousma, pilot. (credit: NASA) S73–31705 (1 Aug. 1973)

Table 3.5 Skylab the first testbed in Space—Things that worked and things that didn't work—Examples (Häuplik-Meusburger 2011, NASA [Bull.10] 1974

Selection of design applications for the Skylab interior	
Crew found applications awkward to use or unnecessary	Crew found design applications useful or pleasant
Chair-type body restraints 'Fireman's pole' for transition Special designed shoes with triangular plates fastened to the soles, to fit with metal triangular grid-work for walls and ceilings In general the interior had many different restraints for the astronauts to test: Velcro restraints, thigh restraints, shoe restraints, temporary restraints, etc. The deployable shower system Inadequate restraint system for food containers and utensils	The half circle dome for 'free exercise' Private sleep compartments Design of the wardroom table, also used for maintenance activities Food system and flatware Windows in the MDA for Earth watching

We should not forget that never before had the astronauts had the possibility to move in such a big volume in microgravity. 'Things that didn't work' included a lot of 0 g restraints that were found not useful or awkward to use. For example, a 'fireman's pole' was not used, and the special design shoes with triangular plates fastened to the soles in order to fit with the metal triangular grid-work of walls and ceilings were not found to be useful (see also Box 3.2).

Box 3.2 Working Around a Poor Design

In a 2017 interview[3] Astronaut Edward Gibson who set an American record for living in space on Skylab of 84 days, related a story of having to make a repair on the Earth Observation antenna which wasn't made to be repaired. The lack of restraints necessitated one astronaut having to wrap his legs and feet around a pole while another astronaut held onto the other to steady him while he attempted to unscrew a critical screw with an undersized screwdriver. Given the bulky suit gloves, this resulted in a job that should have been quickly accomplished taking an hour and a half of hard effort. Upon reentry, Gibson told of how his fingernails were purple from the effort to squeeze the small screwdriver

Paradoxically, while readily acknowledging that living in an extraterrestrial environment could have negative effects and that inappropriate or faulty design could present a threat to the crew's health and the overall mission (Box 3.3), broad application of interior design principles was largely ignored for habitats situated in extreme environments. Before interior design would be seriously considered, the nature of the actual negative effects had to be characterized and determined to be significant enough to warrant substantial changes in the design of proposed extra-terrestrial vehicles and bases as well as those situated in terrestrial ICEs. To complicate matters, most of the research on human adaptation to ICEs[4] or EEs[5] has been focused on individual and group accommodation to the existing austere habitats rather than efforts to improve the habitats themselves. There are only a few examples where designers or architects were involved to improve habitability. Skylab with Raymond Loewy and Salyut and Mir space stations with Galina Balashova were exceptions (see Fig. 3.3a, b) whose innovative visions challenged prevailing engineering attitudes of the day.

Given the low priority of designing for anything except pure functionality, it is no surprise that human experience in 'abnormal' environments reflected abnormal responses when compared to so-called normal environments. However, given the premise that extraterrestrial environments share a core set of limiting characteristics (e.g., microgravity, radiation, no atmosphere) and human habitation in such places

[3]https://gridium.com/space-station-maintenance-skylab/

[4]Isolated and Confined Environments.

[5]Extreme Environments.

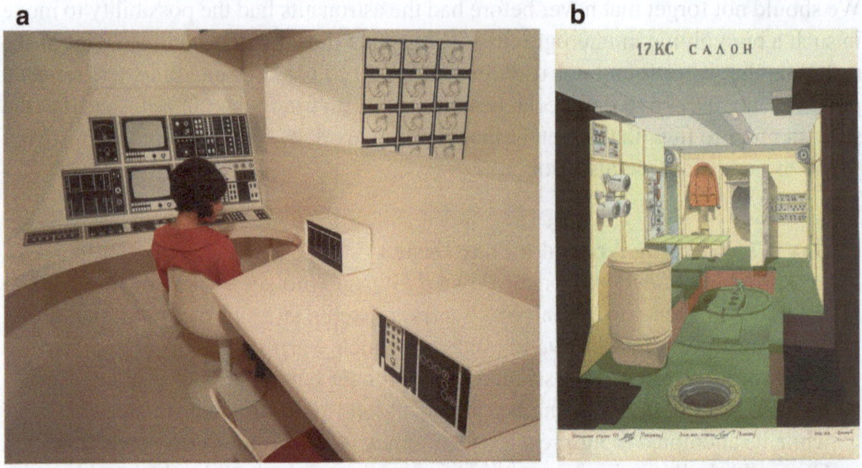

Fig. 3.3 (**a**) Full-scale mockup of a concept for an artificial-G, shuttle-compatible space station interior designed by Raymond Loewy/William Snaith, Inc., for a habitability working group at NASA's Marshall Space Flight Center in Huntsville, Ala (credit: NASA/MSFC). (**b**) Design for the cabinet of the space station Mir, final version interior design (1980) (credit: Archive Galina Balashowa)

share common challenges (e.g., isolation, confinement, restricted access to others), the natural starting point was to look at the commonalities across experience among ICEs and their habitats.

Box 3.3 A Poorly Designed Flexible Air Hose

In 2004 a poorly designed flexible air hose caused a leak at the International Space Station (Oberg 2004). The hose was located in the Destiny science module, close to an optical window for earth-observation (see Fig. 3.4a, b). Due to a lack of appropriate handholds, the astronauts repeatedly held onto the air hose when looking out of the window. This unplanned practice finally resulted in a leaky hose, through which internal air left the station. It has by now been widely acknowledged, that 'window gazing' is the number one leisure activity, and that astronauts and cosmonauts spend a lot of time in front of windows looking at the Earth. Previously Skylab astronaut Gerald Carr stated in 1974, that "*if something is going to stick out and make a nice handhold, it's going to be used for a handhold*" (NASA [Bull.1] 1974, p.76).

Although the consideration of habitability and human factors has been integrated in the design process of human spacecraft, there is still a requirement for improving habitability. This precept is valid also for the design of commercial spacecraft. If designed from a more human-orientated view rather than a solely engineering one, the window would have been provided with means to hold on that was structurally suited for this function.

[As published in the book Architecture for Astronauts (Häuplik-Meusburger 2011, p. 3)].

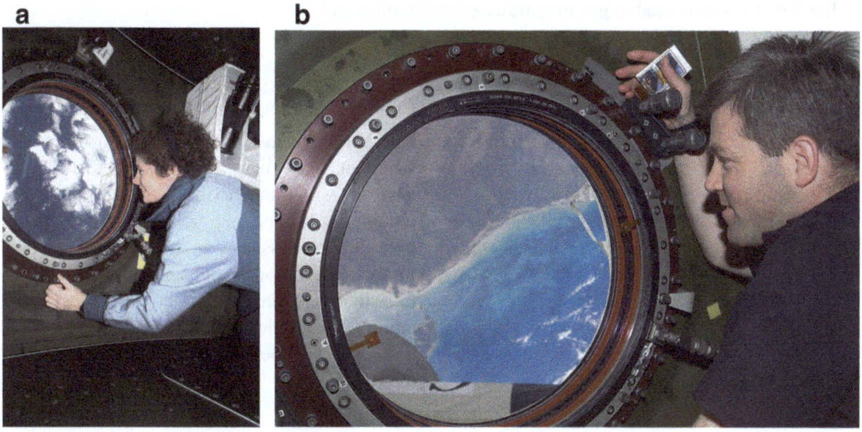

Fig. 3.4 (**a**) Astronaut Susan J. Helms, Expedition Two flight engineer views the topography of a point on Earth from the nadir window in the U.S. Laboratory of the International Space Station. She is holding on the window in the Destiny module, before a handhold was installed and before the incident with the broken hose happened, 2001 (credit: NASA). (**b**) Pilot Stephen N. Frick looks out the nadir window of the U.S. Laboratory/Destiny during STS-110's visit to the International Space Station (ISS). He is holding on the handles that were installed following the incident in 2002 (credit: NASA) (see Box 3.3)

3.2.1 Experience Commonalities Among Extreme Environments

Before we can design living spaces to address limitations imposed by a constrained environment, we first needed to understand the underlying causes of undesirable outcomes. Commonalities in extraterrestrial and isolated confined environments have been detailed in myriad reviews and studies of various extreme environments. Beginning with commonalities in environmental and mission related requirements, anecdotal and then systematic investigation of human experiences in ICEs and space flight environments came under scrutiny. Surprisingly, in early studies, similar individual deterioration in performance, mental functioning, and social interaction was observed in missions as short as two-weeks long and seemed to occur to greater or lesser extent regardless of mission length, danger, and extent of isolation, confinement, nationality or gender (Kanas and Feddersen 1971; Garshnek 1989; Kelly and Kanas 1992, 1993, 1994; Manzey and Lorenz 1997; Herring 1997; Morphew and Maclaren 1997; Kanas et al. 1996; Gushin et al. 1997; Gushin et al. 1996; Kanas 1990, 1998; Kanas and Manzey 2003; Lebedev 1990; Sandal 2004; Bishop 2002). A summary of common challenges can be found in Table 3.6.

Separating individual and group factors from those of the lived environment has been complicated by the unique impact that isolated, confined environments have on participants. The impact of isolation and its sister, confinement, are repeatedly underestimated by participants as well as outside support groups. Little issues can take on atypical importance to teams isolated and confined producing exaggerated

Table 3.6 Common challenges in extraterrestrial habitation

Common challenges in extraterrestrial habitation	
Common challenges to the built structure	Common challenges for the inhabitants
Pressure differential between the pressurized volume and the space environment, microgravity, radiation, extreme temperatures, vacuum or caustic atmosphere, limited possibilities to repair and adapt, abrasive or caustic soils/dust, micro meteors, solar flares	Microgravity, limited space, isolation, confinement, risk, micro-society, sleep disruption, gastrointestinal disturbances, and somatic complaints as well as psychological reports of boredom, restlessness, anxiety, anger, loss of motivation, temporal and spatial disorientation, interpersonal conflict, homesickness, irritability, difficulties in concentration and deficits in task performance over time; monotony

Table 3.7 Crew miscommunication, misunderstanding and interpersonal conflict (Santy et al. 1993)

Mission phase	Number of incidents	Impact		
		Low	Mod.	High
Preflight training	9	6	3	0
Inflight operations	26	9	12	5
Payload/experiments	4	1	3	0
Housekeeping	5	1	3	1
Personal hygiene	5	1	3	1
Postflight activities	7	3	4	0
Totals	42	18	19	5

responses to stimuli (or lack of) that would not normally provoke such responses. For instance, one of the first surveys of American astronauts who flew on international space missions between 1981 and 1990 conducted by Santy et al. (1993) found the highest rates of miscommunication, misunderstandings and interpersonal conflicts were in the moderate to high mission impact category and occurred during the in-flight operations over mundane and routine matters involving payload/experiment resources, housekeeping tasks and personal hygiene issues (see Table 3.7).

Anecdotal evidence from the earliest extreme environments similarly clearly indicated problematic areas in which even short-term individual and group functioning had been compromised to some extent by the presence of communication breakdowns, interpersonal conflict, individualized responses to environmental stressors and conflicts over authority and control (Anderson 1991; Bishop et al. 1999; Herring 1997; Kanas et al. 2001a, b; Morphew and Maclaren 1997; Sandal et al. 1996; Santy 1994; Santy et al. 1993). In particular the following problems have been reported on space stations, submarines, polar and simulator facilities: Interpersonal conflicts, somatic complaints, sleep disturbances, boredom, restlessness, decrements in performance, and decline in group compatibility. In many of them substance abuse has been confirmed.

The well-documented **phenomena of displacement**[6] and development of hostility towards 'mission control' and 'home office' groups is at least partially due to a lack of shared perspective as the lived experience of crews grows increasingly more discrepant from those on the ground (Nicholas 1987; Kanas et al. 2001a, b) (Box 3.4). The very fact that these phenomena have occurred in almost all ICE groups, regardless of specific environment, and to varying extent to most individuals involved regardless of culture or gender strongly suggests that future extraterrestrial groups will be at risk for these experiences as well.

Box 3.4 Astronauts on Strike

Gerald P. Carr, Edward G. Gibson, and William R. Pogue were the third and last crew of the Skylab space station (Skylab 4, also SL-4). On December 28, 1973, for a little over 90 minutes, or one full orbit of Earth, NASA mission control in Houston, Texas, lost contact with the crew. Some have called this the first strike in space, due to the heavy workload of the astronauts, and the astronauts asking for a Sunday off. This is what Ed Gibson says about this day:

> He [Gerald 'Jerry' Carr] was trying to be a good commander and he was watching out for us, because he could see that we were really getting worn down to a little nubby trying to work as late as we could and getting up early and just trying to make the whole thing work. So he said, "Look. This is not the way. We need Sundays off." So all of a sudden it was "We're going on strike," and that somehow got out in the press. So I still hear about it today. Because Jerry asked for Sundays off. I was going to work Sundays anyway, because I always did. I always worked every day. Every minute I was up there, I was doing something.
>
> [...] What happened was that the ground got to be a little obnoxious at times, just continually asking for one thing after another. And every time we'd come up on a ground station, we'd start working, you'd have to drop whatever you were doing and go on over and talk to them. We said, "Well, let's make it only so that one person has got to do that. We'll take turns. The rest of the guys, you turn off your radios and just one person does it."
>
> Well, we screwed up again. We ended up in a situation where all of our radios were off. We didn't have it right who was doing what. So there was about an orbit went by and finally we said, "We haven't heard anything from the ground, have we." So then we turned the radio back on. Of course, they'd been calling us all that time, and they attributed it that it was something deliberate. At least the press did, anyway. (NASA Johnson Space Center Oral History Project, Edited Oral History Transcript 2000)

As it became undeniable that many ICE psychosocial stressors can and did lead to degraded performance, the contribution of impoverished habitability went through a similar cycle of tacit recognition and similar disregard. As early as 1985, commonly reported habitat related experiences were "*problems associated with interior space, food, hygiene, temperature, decor, odor, and noise*" (Connors et al. 1985, p. 60). In

[6]Displacement is a psychological defense mechanism in which negative feelings are transferred from the original source of the emotion to a less threatening person or object (see Kanas and Feddersen 1971).

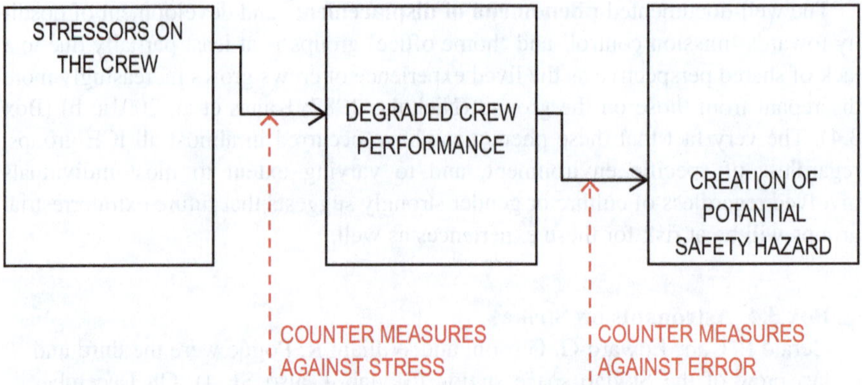

Fig. 3.5 The crew safety-human factors interaction model (adapted from Cohen and Häuplik-Meusburger 2015; original image by Marc M. Cohen)

Table 3.8 Stressors for Long-term human spaceflight and possible countermeasures related to space architecture (Dudley-Rowley et al. 2004)

Stressors associated to habitability	Effect on human behavior	Architectural countermeasures against stress
Volume limitations	Lack of privacy; feelings of claustrophobia	Interior layout, interior zoning, windows (real and virtual), integration of technology (VR, MR)
Confinement; isolation; separation	Feelings of claustrophobia; lack of motivation; "cabin fever"; degraded performance	Layout (social events, visitors, private communication with family and friends)
Noise, vibrations	Sleep disturbances, poor communication and misunderstandings	Vibration, isolation and control, spatial and temporal zoning of activities
Lighting, illumination	Fatigue; irritability; blurred vision	Lighting design, mix of natural light, ambient light and task-specific lighting

the same year, the Space Station Crew Safety Alternatives Study (Cohen and Junge 1984) introduced the **Crew Safety-Human Factors Interaction Model**. The relationships and direct influences are illustrated in Fig. 3.5. "*In this model, a stressor (such as one of the threats) can lead to degraded performance, which can contribute to human error, unless appropriate and effective countermeasures are available to the crew*." (Dudley-Rowley et al. 2001, p. 1).

In 2004, M. Dudley-Rowley, M. M. Cohen and P. Flores conducted a comparison of the 1985 NASA study with the safety record on Mir (operational between 1986–1996). Their findings highlighted the relationship between stressors and the architecture of the habitat and identified examples of stressors, their effects, as well as suitable architectural and design countermeasures (see Table 3.8). The overlap between negative outcomes from psychosocial stressors and those related to poor habitability argued strongly for treating these as interactive factors.

It wasn't until more analogue habitats became available and sample sizes accrued across environments that it became possible to begin a closer examination of how habitability affected human performance and wellbeing in ICEs amidst the noise of uncontrolled, highly variable and volatile environments populated by a hugely heterogeneous pool of participants. Across time, analogues, and studies, most of the long duration performance effects found so far seem to be associated with general stress effects related to problems of **adaptation to the extreme living and working conditions in a confined and isolated environment** which are mediated by individual factors such as personality and culture and habitability factors. For example, whether lack of privacy is perceived as a stressor, and thus produces detrimental effects on mood and performance, largely depends on the cultural background of individuals (Raybeck 1991) and their embedded fit to the situation (living space in which they find themselves). All ICEs are characterized by very limited personal quarters which subsequently impact perceptions of privacy, density and crowding. Some cultures are far more comfortable with small physical living spaces than others. Women have been found to tolerate higher density better than males yet prefer greater degrees of privacy.

Privacy is also connected to the overall mission scenario and layout of the habitat (see Sect. 7.7.1). During the Apollo missions, astronauts slept, shaved and defecated next to each other in one volume and had no privacy, but since "*most pilots are not used to privacy* (...) *it was not a problem*"[7]. The Salyut cosmonauts did not have private crew quarters. They slept next to each other and preferred having a visual overview of the station, which made them feel more secure. The Skylab station and the MIR space station provided centrally located but spatially enclosed crew quarters. Both stations eventually consisted of various modules; this offered the opportunity for astronauts to relocate their place to sleep in another (not dedicated) part of the station.

So, the same living quarters have been very differently experienced across individuals depending on a multitude of mediating factors! Similar effects have been found as well as other stressors, like monotony and boredom (NRC, Space Studies Board 1998, pp. 94–227; Kanas 1998; Otto 2007; Basner et al. 2014; Vessel and Russo 2015), sleep and circadian rhythms (NCR, Space Studies Board NRC, Space Studies Board 1998; NASA 1991; Lim and Dinges 2010; Barger et al. 2014), or autonomy and meaningful work (Vanhove et al. 2014). Poor habitability not only appears to contribute to ICE stress and subsequent poor adaptation but several recent lines of research have suggested that architectural design, itself, may independently contribute to poor (or otherwise better) adaptation (Kearney 2013; Stuster 1996; Kanas and Manzey 2008a, b).

[7]Schmitt, Harrison. 2009. Transcript not published. [interv.] Sandra Häuplik-Meusburger. Vienna, Austria, 2009.

Table 3.9 Summary of recent and operational short-term and long-term studies related to Habitability Research

Environment	Short-term missions <90 days	Long-term missions >90 days
Space	[a]Shuttle [a]Space stations (Salyut, Mir, Skylab, ISS)	[a]Space Stations (Mir, ISS)
Earth-bound Site	[a]McMurdo, South Pole Station summer crews [a]Nuclear submarines (3–4 m) [b]Aquarius/NEEMO (10d) [b]MDRS (14d) [c]HERA	[a]McMurdo, South Pole Station winter over crews [a]Lunar Palace (Youegong-1) [b]Biosphere-2 (700d) [b]Concordia (420d) [b]HI-SEAS [c]IBMP/SFINCSS'99/Mars500

[a]Operational facilities
[b]In-situ simulators
[c]Behavioral laboratories

3.2.2 *Move from Short Duration Missions to Long Duration Missions*

Once the realization that the interaction between mission and human related characteristics and human responses were far more critical in defining effective living spaces for extreme environments, the logical evolution was to move away from trying to utilize short duration transition spaces (e.g., Shuttle, MDRS, HERA) as models for long duration living spaces. Consequently, the shift from short duration exposure studies to studies of long duration human adaptation to ICE environments also provided an opportunity to more closely examine the importance of factors that supported thriving rather than merely surviving. This required a shift from the importance of *place* to importance of living *space*, i.e., instead of the characteristics of the location being primary design drivers, the question was asked how the space provided satisfied the needs of the inhabitants living within given the mission they were tasked to complete. Short duration living spaces were effectively relegated to proof of concept and hardware testbeds while analogues characterized by long duration mission profiles rose in desirability. The result was a lengthening of mission duration in many simulation facilities in addition to expanding research into existing operational environments to achieve the desired mission profile (see Table 3.9).

However, identifying the relevant characteristics of long duration habitats has not been a straightforward task (Box 3.5). Many ICE analogue environments have non-research mandates that still prioritize operational functionality (e.g., Antarctic bases, military submarines) over testing provisions for optimal human health and wellbeing. Even in pure research analogues, there have been various attempts to create habitats that incorporated key characteristics that were subsequently undermined by intervention. For example, there was the famous problem in Biosphere 2 with the absorption of oxygen by microbes in the soil occurring at a higher rate than expected. This resulted in higher levels of CO_2 that should have stimulated compensatory plant growth to replace the consumed oxygen but instead was absorbed by the concrete structure. At 18 months into the 2-year mission, it was necessary to provide supplementary oxygen in order to finish the project (MacCallum et al. 2004). Had this been on Mars or a space station, a very different solution involving changes in structure would have been necessitated (see also Box 7.4 in Chap. 7).

Box 3.5 What Constitutes a Long Duration Mission?

Deciding on what kind of habitats and situations qualified for inclusion in this book has been an interesting exercise. Take sailing ships for instance. Like space missions, they are expeditions, i.e., move from place to place in a progressive sequence but they are also habitats since a persistent shelter is maintained. They sometimes entail long continuous periods of isolation and confinement. The longest ocean voyage was apparently achieved by Reid Stowe in 2010 who deliberately spent 1152 days sailing a 70 ft. (21.3 m), 60-ton (54,400 kg) gaff-rigged schooner around the ocean without resupply or stopping in a harbor simulating a Mars mission duration (https://en.wikipedia.org/wiki/Reid_Stowe). If we use time alone as a defining metric, this 'mission' would qualify. But that was one guy and he had his girlfriend with him the first quarter of the journey. There are other long cruises but they all stop frequently. And early explorers stopped at every available bit of land they found and were made up of crews significantly larger than most projected space crews. However, we decided to include nuclear submarines on the list because of their extreme isolation protocols but not sailing ships. *At some point, the distinction between classifications can go either way.*" Sheryl Bishop, Jan. 2021

3.3 Habitability Design as Countermeasure: Addressing the Negative Effects of ICEs

Addressing the negative impacts of confinement and isolation with habitability factors has been advocated for decades in various planning workshops for both the moon and Mars (Connors et al. 1985; Stuster 1989; Stuster et al. 2000). It has been argued that physical and operational environmental variables that significantly impact individual and group behavior can be mitigated by 'state of the art' habitability and facilitate human welfare and successful performance during long duration missions. The 'risk of incompatible vehicle or habitat design' has been identified by NASA as a recognized risk to human health and performance in space (Whitmore et al. 2013, p. 3).

Many have outlined detailed lists of critical features that needed addressing (see Table 3.10). In 1985 the NASA report 'Living Aloft' (Connors et al. 1985) described a number of *projected* habitability issues of extended spaceflight. Twenty-five years later, in 2010, Stuster identified a list of 24 issues with behavioral implications for human spaceflight derived from a content analysis of personal NASA astronaut journals that were maintained for this purpose during expeditions onboard the International Space Station. According to Stuster, the study provided the first

Table 3.10 Habitability design

Relevant issues potentially addressed through design
The Physical Environment (Interior Space, Food, Hygiene, Temperature and Humidity, Décor and Lighting, Odor, Noise), Health and Leisure (Recreation, Exercise), Privacy (Crowding, Territoriality), Complex Effects. Connors et al. (1985)
Including habitable volume, crew quarters, leisure applications, décor, and windows. Kanas and Manzey (2003)
Outside Communication, Group Interaction, Recreation/Leisure, Sleep, Food (Top 10 of 24 issues); Clothing, Exercise, Medical Support, Personal Hygiene, Habitat Aesthetics, Privacy Personal Space, Waste Disposal and management, Onboard training, Simulation and task preparation. Stuster (2010)
Sleep (rest, relaxation, sleep and storage), Hygiene (personal hygiene, shower, toilet, housekeeping), Food (store, prepare, grow, consume, and storage), Work (operations, experiments, communication, education, training, and storage), Leisure (free-time activities, exercise, intimate behavior, and storage). Häuplik-Meusburger (2011)

Relevant Issues identified in Human Factors Research, that can be potentially addressed through design and architecture (Sources: As in table)

quantitative data on which he based a rank-ordering of the behavioral issues associated with long duration space operations (Stuster 2010). The top 10 issues account for 88% of all astronaut journal entries that he had examined and are listed in along with other behavioral and habitability issues identified by various researchers in human factors research. From the list of 24 issues, Stuster further identified 14 behavioral issues (see Table 3.10) that could potentially be addressed through habitat design: sleep, clothing, exercise, medical support, personal hygiene, food preparation, group interaction, habitat aesthetics, outside communications, recreational opportunities, privacy and personal space, waste disposal and management, onboard training, simulation and task preparation, and behavioral issues associated with the microgravity environment.

Habitability design can be envisioned as a viable contributor to both active and passive countermeasures for certain stressors. Particularly for long, remote mission scenarios, mission stress can be reduced through internal architecture and systems (Winisdoerffer and Soulez-Larivière 1992, p. 315; Bishop et al. 2016). Examples of areas where appropriate design has already been employed and acknowledged as a countermeasure are listed in Table 3.11.

3.4 How to Evaluate and Prove Effectiveness: Operational Versus Research Simulation Facilities

So, how are we to evaluate and prove effectiveness of habitation factors when we lack true analogues for extraterrestrial conditions? Why don't we know more about what is needed? As long duration missions are most relevant to extraterrestrial habitability issues, scrutiny of existing long duration facilities highlight why advancement of habitability research in ICEs has been so agonizingly slow. Progress

Table 3.11 Examples of effective design countermeasures

Goal of the (counter) measure	Problem	Exemplary means of how to achieve the goal
Enhancing performance	Artificial and inadequate lighting can lead to fatigue, irritability and blurred vision. A potential safety hazard is mistaken perception	Appropriate lighting design can counteract degraded performance
Enhancing psychological functioning	Constant confinement and isolation as well as the decrease of privacy can lead to feelings of claustrophobia, loneliness and impaired judgment	Adaptable interior configurations allow for social group activities and changes in an environment characterized by monotony and overfamiliarity
Enhancing social cohesion	The withdrawal from the normal social matrix and dependence on a small community can lead to depressed mood, social withdrawal or group splintering	Appropriate habitat layout can facilitate social interaction (e.g., events, group gatherings, shared activities) and as well as provide for private interactions (e.g., communications with family and friends, small groups/dyads) which can counteract the negative effects of isolated, confined environments

has been slow primarily due to the overriding focus on survivability (see Chap. 2, and Box 3.1 as an example for the challenging communication on habitability issues). Existing facilities currently vary widely in their existing structures as well as their purpose. For instance, in Antarctica, acknowledged as the 'best' analogue environment for long duration space missions due to the extensive duration of winter isolation, there exists only one pure research facility, Concordia (see also Sect. 5.1.2.3), a collaboration between France and Italy, consisting of three buildings, which are interlinked by enclosed walkways, built on Dome C, one of the coldest and most remote places on Earth. Operational since 2005, the station can accommodate 16 persons during the winter and 32 people during the summer season. Participants are selected via an involved process of screening and selection procedures. Still, "*How to pack your mind for a trip away from home, from friends and family, for a period of a year?*" That's one of the most challenges voiced by Alexander Kumar while wintering over at Concordia in Antarctica (Kumar 2012) (Box 3.6).

Box 3.6 A Home in Antarctica

In an interview, medical doctor and researcher Alexander Kumar, was asked which colors and sounds he associates with his home now (while being in Antarctica). This is what he answered: *I will give you an example. One of the glaciologists here caused a leak, a flood of water, that leaked from the top floor to the middle floor. When I opened the door and walked out, it was the*

(continued)

Box 3.6 (continued)
sound of raindrops. And that reminded me of the North of England, where I grew up. Because it always rains there. So, in terms of sounds, tastes and smells, I would say that all senses, where you [me] live have color, but here everything becomes black and white. There is an element of sensorial deprivation of course, but, you know, you have to learn how to get the best . . . for example if you have someone that has partially or totally loss of sight, you have to understand how to get the best from your hearing, if you can't see. And vice versa. The interview with medical doctor and researcher Alexander Kumar was conducted in 2012, while he was in Antarctica via Skype (Kumar 2012). Listen to the interview: https://cba.fro.at/65695

All other bases in Antarctica would be considered operational environments supporting scientific, logistic, strategic and military programs. The largest is McMurdo Station (Fig. 3.6), operated by the United States on Ross Island. Supporting a population of 1000 in the summer months and approximately 200 in the winter months, the site is a modern-day operational science station, which includes a harbor, three airfields (two seasonal), a heliport and more than 100 buildings. Functionally, it is a base with seasonal and transient (albeit some with considerable tenure) inhabitants and not a settlement. Participants are primarily screened for physical and psychiatric disqualifying conditions with an emphasis on skill set but with little attention given to compatibility or fit within the larger group (see Sects. 7.7 and 7.4.1).

At the other end of the operational base spectrum is Chile's Presidente Eduardo Frei Montalva Base which includes a residential area with a hospital, school, bank, small supermarket, etc. The maximum population during summer is 150 people and about 80 people during the winter, including families with children (https://en.wikipedia.org/wiki/Base_Presidente_Eduardo_Frei_Montalva, accessed 4/8/21). Participants are assigned for multi-year rotations and the sense of community is prominent.

While situated in true extreme environments posing real risks, ICE operational long duration facilities like McMurdo Station are compromised by their focus on non-research priorities and difficulty in implementing scientific protocols to test human adaptation or station modification to test habitation configurations. Laboratory simulation chamber facilities (e.g., HERA, IBMP) lack real danger and risk. A compromise eventually emerged that took the form of habitats built specifically for research purposes situated in moderately challenging or isolated locations in which more experimental control could be implemented as well as moderate confinement and isolation. Chamber facilities and the in-situ simulation habitats utilize minimalist spartan chambered structures in which to simulate the confinement and limited isolation of missions of varying durations. Like the International Space Station, there is almost no accommodation beyond functionality. Unlike the ISS, these facilities only loosely mimic protocols for EVA egress/ingress, communication delays, the employment of space suits and other factors that would be in place in

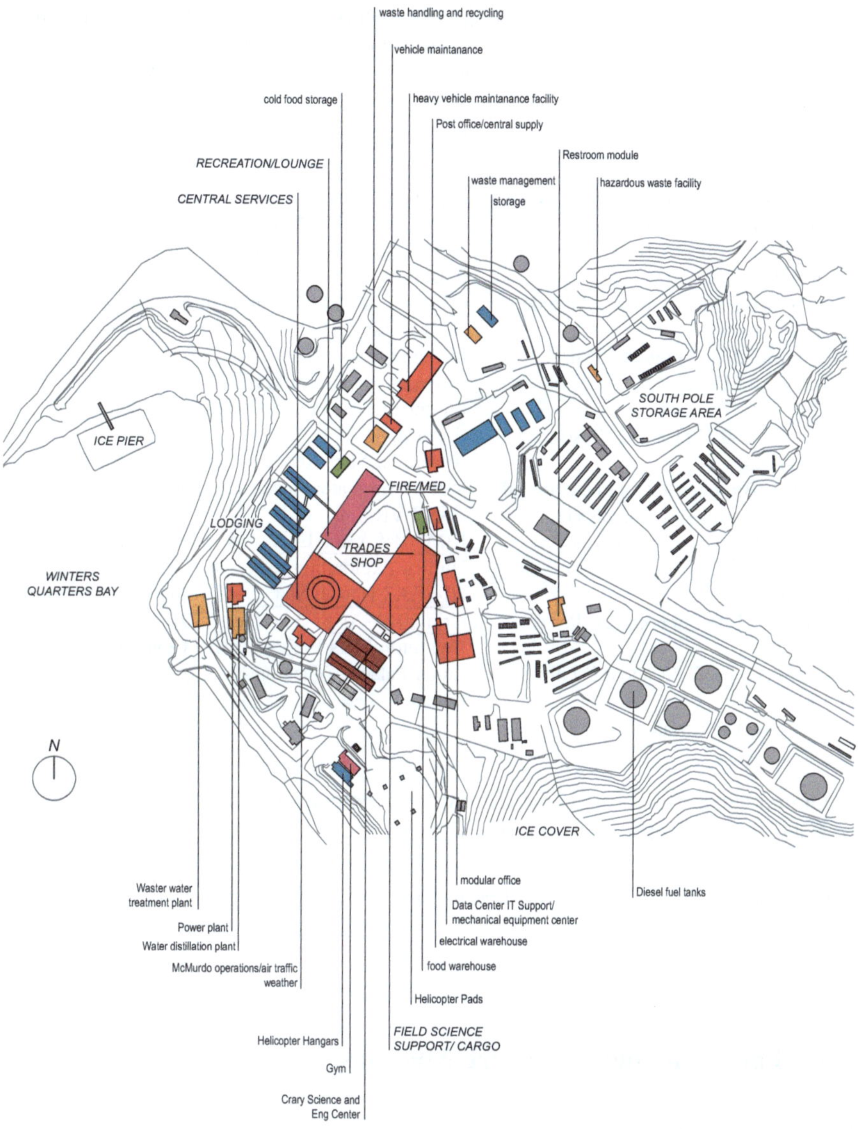

Fig. 3.6 Overview Site plan of the McMurdo Station (credit: The Authors)

an actual extraterrestrial environment. Most are not situated in places characterized by significant environmental risk as with Antarctica. They do, however, provide the opportunity to be somewhat modified in order to target specific design parameters of interest and, thus, offer opportunities for assessing efficacy and effectiveness of different design implementations. Efforts to test various components of habitability and human adaptation in these modular structures has generated growing interest in the construction of new research facilities that can be more easily modified to

Table 3.12 Exemplars of past and current chamber facilities and simulation habitats. The facilities simulating long-term human space missions exceeding 6 months are underlined and those in realistic environments are noted with an asterisk. (updated list from Häuplik-Meusburger et al. 2017)

Past simulators and simulation missions	Current simulators & simulation missions (as of 2021)
Regenerative Life Support Study by NASA Langley Research Center	Aquarius and NASA Extreme Environment Mission Operations (NEEMO)*
Apollo Ground-based Tests	Mars Desert Research Station (MDRS)*
MOLAB	Flashline Mars Arctic Research Station (FMARS)*
Skylab Medical Experiments Altitude Test (SMEAT)*	Concordia Research Station in Antarctica*
Skylab Mobile Laboratory (SML)*	NASA Fast Track Horizontal and Vertical Mock-Ups for lunar habitation
Ben Franklin Underwater Research Laboratory*	Environmental Habitat (EnviHab)
Tektite I and II Underwater Research Laboratories*	European Mars Analog Research Station (EuroMARS)*
BIO-Plex (Bioregenerative Planetary Life Support Systems Test Complex)	Australian Mars Research Station (MARS-Oz)*
BIOS-3 (Institute of Biophysics, Krasnoyarsk, Russia)	Virtual Simulators located at Industries, such as TAS-I VR Lab
Biosphere-2*	IBMP (RSA and ESA)
Lunar Mars Life Support Test Project (LMLSTP)	Human Exploration Research Analog (HERA)
Closed Ecology Experiment Facilities (CEEF)	HI-SEAS Hawaii Space Exploration Analog (long-term 2013–2018*; short-term 2018-ongoing)
Jules Verne Underwater Facility*	Haughton Mars Project (HMP) Devon Island
	Lunar Palace—Yuegong-1
	Lunares

accommodate a wider range of interior design elements under conditions of varying complexity to test efficacy and effectiveness.

Table 3.12 displays a list of past and present analogue habitats that have incorporated habitability features and begun to approach full-scale simulation research facilities.

3.5 The Challenge of Future Missions

In the context of future extraterrestrial missions, it may be that the most severe stressors may involve monotony and boredom resulting from low workload, hypostimulation, and restricted social contacts due to extreme isolation from family and friends, as well as new forms of social relations requiring increased spatial flexibility. The challenge is that current countermeasure strategies developed for near-Earth orbital facilities will be less effective or completely inoperable.

For instance, support strategies that currently provide ISS crews private real-time family conferences and internet phone facilities will not be possible during longer missions, e.g., to Mars, where the communication lag is 20 min or more one-way. Similar current efficient strategies to ameliorate stress levels during short-term missions (e.g., surprise packages or fresh foodstuffs on resupply vessels) will be impossible to provide during interplanetary voyages and Mars missions.

Individual coping and adaptation is strongly related to various interpersonal factors; however, leadership style and group dynamics are also key factors responsible for exacerbating or ameliorating stress, or facilitating coping and adaptation (Kanas 1997; Kanas and Manzey 2003; Sandal et al. 1995). Currently, all missions are heavily overseen, managed and monitored by respective national mission control organizations. Not only will the burden of support have to shift away from the 'absent' network of family and friends to members of the present crew, operational control will involve significantly higher degrees of (up to 100%) autonomy and local decision making across the board. Crew members will have to be able to balance needs for affiliation and support with autonomy and privacy needs. Individuals with high needs for social support will likely be more susceptible to the negative effects of social isolation from friends and family (Sandal 1999) as well as more likely to engage in interpersonal disclosure that later leads to discomfort and regret (Suedfeld 2003). Selected individuals will be characterized by effective coping strategies based on self-reliance and autonomy as well as cooperative group focus.

A number of habitability factors will impact the coping ability of crewmembers in dealing with these new and challenging living conditions (Kanas 1997; Kanas and Manzey 2003; Sandal et al. 1995). Both interior habitability design integration and mission structure factors will play a significant role in generating group fusion (e.g., cohesion, teamwork, interpersonal trust) or group fission (e.g., discord, conflict, miscommunication, reduction in performance and wellbeing). In this context, the layout of the habitat itself, and its spatial adaptability and flexibility will probably play a much more important role.

An effective extraterrestrial habitat needs to actively provide for a multitude of needs and demands that would otherwise be supplemented on Earth through other sources. ***What you take is what you have*** in terms of people, space and equipment. Everything has value. All elements will interact with each other in a much more open way. Thus, space habitats need to more holistically encompass the spectrum of human existence rather than simply the segment we spend when not at work or elsewhere. There is no 'elsewhere' to be.

In Chaps. 4 and 5, relevant research findings addressing habitability issues for these facilities will be covered.

References

Anderson, C. (1991). Polar psychology – Coping with it all. *Nature, 350*(6316), 290.

Barger, L. K., Flynn-Evans, E. E., Kubey, A., et al. (2014). Prevalence of sleep deficiency and use of hypnotic drugs in astronauts before, during, and after spaceflight: An observational study. *The Lancet. Neurology, 13*(9), 904–912. https://doi.org/10.1016/S1474-4422(14)70122-XI.

Basner, M., Dinges, D. F., Mollicone, D. J., Savelev, I., et al. (2014). Psychological and behavioral changes during confinement in a 520-day simulated interplanetary Mission to Mars. *PLoS One, 9*(3), e93298.

Bishop, S. (2002). Evaluating teams in extreme environments: Deep caving, polar and desert expeditions, *Society of Automotive Engineers*. In *32nd international conference on environmental systems (ICES)*, San Antonio, Texas, USA, July 2002.

Bishop, S. L. (2006). Psychological and psychosocial health and well-being at Pole Station. In C. S. Cockell (Ed.), *Project Boreas: A station for the Martian geographic North pole* (pp. 160–171). London: British Interplanetary Society.

Bishop, S. L., Santy, P. A., & Faulk, D. (1999) Team dynamics analysis of the Huautla cave diving expedition: A case study, 1999-01-2099. *SAE Technical Papers*. https://doi.org/10.4271/1999-01-2099

Bishop, S., Haueplik-Meusburger, S., Guined, J., & Peldszus, R. (2016). *Bionomic design countermeasures for enhancing cognitive and psychological functioning and crew performance in isolated and confined habitats*. International Congress for Environmental Systems, Space Architecture, Vienna, Austria, July 10–14, 2016.

Cohen, M. (2012). *Mockups 101: Code and standard research for space habitat analogues*. AIAA Space 2012 Conference. Pasadena, California, 2012.

Cohen, M. M., & Häuplik-Meusburger, S. (2015). *What do we give up and leave behind?* ICES-2015-56, 45th international conference on environmental systems, Bellevue, Washington, 12–16 July 2015.

Cohen, M. M., & Junge, M. K. (1984). Space station crew safety: Human factors model. *Proceedings of the Human Factors Society Annual Meeting, 28*(10), 908–912.

Compton, W. D., & Benson, C. D. (1983). *Living and working in space, a history of Skylab*. Washington, DC: National Aeronautics and Space Administration.

Connolly, J., & Daues, K., & Howard, Jr., R. L., & Toups, L. (2006). Engineering, construction, and operations in challenging environment. In *Definition and development of habitation readiness level (HRLs) for planetary surface habitats, earth & space* (pp. 1–8). Reston, VA: American Society of Civil Engineers.

Connors, M. M., Harrison, A. A., & Akins, F. R. (1985). *Living aloft. Human requirements for extended spaceflight*. Washington, DC: NASA Scientific and Technical Information Branch.

Dudley-Rowley, M., Whitney, S., Bishop, S., Caldwell, B., & Nolan, P. D. (2001). *Size, composition and time: Implications for habitat and workplace design in extreme environments*. Accessed July 1, 2020, from https://www.researchgate.net/publication/272092423_Crew_Size_Composition_and_Time_Implications_for_Habitat_and_Workplace_Design_in_Extreme_Environments

Dudley-Rowley, M., Cohen, M. M., & Flores, P. (2004). 1985 NASA-Rockwell Space Station crew safety study: Results from Mir. *Aviakosmicheskaia i Ekologicheskaia Meditsina, 38*(1), 15–28.

Garshnek, V. (1989). Crucial factor: Human. *Space Policy, 5*(3), 201–216. https://doi.org/10.1016/0265-9646(89)90087-8.

Gushin, V. I., Kolintchenko, V. A., Efimov, V. A., & Davies, C. (1996). Psychological evaluation and support during EXEMSI. In S. Bonting (Ed.), *Advances in space biology and medicine* (pp. 283–295). London: JAI Press.

Gushin, V. I., Zaprisa, N. S., Kolinitchenko, T. B., Efimov, V. A., Smirnova, T. M., Vinokhodova, A. G., & Kanas, N. (1997). Content analysis of the crew communication with external communicants under prolonged isolation. *Aviation, Space, and Environmental Medicine, 68* (12), 1093–1098.

Häuplik-Meusburger, S. (2011). *Architecture for astronauts: An activity-based approach*. Springer Science & Business Media.

Häuplik-Meusburger, S., & Bannova, O. (2016). *Space architecture education for engineers and architects*. Springer International.

Häuplik-Meusburger, S., Binsted, K., & Bassingthwaighte, T. (2017). *Habitability studies and full scale simulators: Preliminary themes following HISEAS mission IV*. In 47th international conference on environmental systems, Charleston, South Carolina, 16–20 July 2017.

Herring, L. (1997). Astronaut draws attention to psychology, communication. *The Journal of Human Performance in Extreme Environments, 2*(1), 42–47.

Kanas, N. (1990). Psychological, psychiatric, and interpersonal aspects of long-duration space missions. *Journal of Spacecraft and Rockets, 27*(5), 457–463. https://doi.org/10.2514/3.26165.

Kanas, N. (1997). Psychosocial value of space simulation for extended spaceflight. In S. L. Bonting (Ed.) Advances in Space Biology and Medicine, V6, pp 81–91.

Kanas, N. (1998). Psychiatric issues affecting long duration space missions. *Aviation, Space and Environmental Medicine, 69*(12), 1211–1216.

Kanas, N., & Feddersen, W. E. (1971). *Behavioral, psychiatric, and sociological problems of long duration space missions. NASA TM 58067*. Houston, TX: NASA Johnson Space Center.

Kanas, N., & Manzey, D. (2003). *Space psychology and psychiatry*. Dordrecht: Springer/Microcosm Press.

Kanas, N., & Manzey, D. (2008a). *Space psychology and psychiatry* (2nd ed.). El Segundo, CA: Microcosm Press.

Kanas, N., & Manzey, D. (2008b). Basic issues of human adaptation to space flight. In *Space psychology and psychiatry* (pp. 15–48). Dordrecht: Springer.

Kanas, N., Weiss, D. S., & Marmar, C. R. (1996). Crewmember interactions during a Mir space station simulation. *Aviation, Space, and Environmental Medicine, 67*(10), 969–975.

Kanas, N., Salnitskiy, V., Grund, E. M., et al. (2001a). Psychosocial issues in space: Results from shuttle/Mir. *Gravitational and Space Biology Bulletin, 14*(2), 35–45.

Kanas, N., Salnitsky, V., Weiss, D. S., et al. (2001b). Crewmember and ground personnel interactions over time during shuttle/Mir space missions. *Aviation, Space and Environmental Medicine, 72*, 453–61.16.

Kearney, A. B. (2013). *Mitigating risks to psychological and behavioral health in long duration spaceflight: Towards a portfolio of countermeasures* (White paper). Prepared for NASA-Johnson Space Center, Houston, TX: Behavioral Health and Performance.

Kelly, A. D., & Kanas, N. (1992). Crewmember communication in space: A survey of astronauts and cosmonauts. *Aviation Space and Environmental Medicine, 63*(8), 721–726.

Kelly, A. D., & Kanas, N. (1993). Communication between space crews and ground personnel: A survey of astronauts and cosmonauts. *Aviation Space and Environmental Medicine, 64*(9 Pt 1), 795–800.

Kelly, A. D., & Kanas, N. (1994). Leisure activities in space – A survey of astronauts and cosmonauts. *Acta Astronautica, 32*(6), 451–457.

Kumar 2012, interview, radio broadcast, Häuplik-Meusburger, S., Holzgethan, V. (Host). (2012, October 9). *Alexander Kumar gives an insight to daily life in Antarctica*. In Spacecuriosity. https://cba.fro.at/65695

Lebedev, V. (1990). *Diary of a cosmonaut – 211 days in space*. United States and Canada, Bantam Air & Space Series. ISBN: 0-553-28778-8.

Lim, J., & Dinges, D. F. (2010). A meta-analysis of the impact of short-term sleep deprivation on cognitive variables. *Psychological Bulletin, 136*(3), 375–389. https://doi.org/10.1037/a0018883.

MacCallum, T., Poynter, J., & Bearden, D. (2004). Lessons learned from biosphere 2: When viewed as a ground simulation/analog for long duration human space exploration and settlement. *SAE Transactions, 113*(1), 1095–1104.

Manzey, D., & Lorenz, B. (1997). Human performance during prolonged space flight. *Journal of Human Performance in Extreme Environments, 1*(2), 68.

Manzey, D., Schiewe, A., & Fassbender, C. (1995). Psychological countermeasures for extended manned spaceflights. *Acta Astronautica, 35*(4–5), 339–361.

Morphew, M. E., & Maclaren, S. (1997). Blaha suggests need for future research on the effects of isolation and confinement. *Journal of Human Performance in Extreme Environments, 2*(1), 52–53.

NASA. (1991). *Space human factors discipline science plan*. Space Human Factors Program, Life Sciences Division. NASA Headquarters, Washington, DC. Retrieved May 18, 2020, from https://ntrs.nasa.gov/archive/nasa/casi.ntrs.nasa.gov/19930012181.pdf

NASA [Bull.1] (1974). *Skylab Experience Bulletin No.1: Translation Modes and Bump Protection*. Houston, Texas: NASA, Lyndon B. Johnson Space Center, 1974. JSC-09535.

NASA, Johnson Space Center Oral History Project, Edited Oral History Transcript. (2000). Edward G. Gibson, Interviewed by Carol Butler, Houston, Texas – 1 December 2000. Retrieved May

18, 2020, from https://historycollection.jsc.nasa.gov/JSCHistoryPortal/history/oral_histories/GibsonEG/GibsonEG_12-1-00.htm

Nicholas, J. M. (1987). Small groups in orbit: Group interaction and crew performance on Space Station. *Aviation, Space and Environmental Medicine, 58*, 1009–1013.

NRC, Space Studies Board. (1998). *A strategy for research in space biology and medicine in the new century. Behavioral issues.* Washington, DC: National Academy Press.

Oberg, J. (2004). *Space station crew prepares to fix window: Damage accidentally caused last month.* NBC News. Retrieved September 2, 2020, from http://www.nbcnews.com/id/4253674/ns/technology_and_science-space/t/space-station-crew-prepares-fix-window/#.X0_XH8hKiUk

Otto, C. A. (2007). *Antarctica: Analog for spaceflight.* Presentation to NASA BHP. Houston, TX, February 2007.

Raybeck, D. (1991). Proxemics and privacy: Managing the problems of life in confined environments. In A. A. Harrison, Y. A. Clearwater, & C. P. McKay (Eds.), *From Antarctica to outer space* (pp. 317–333). New York, NY: Springer.

Sandal, G. M. (1999). The effects of personality and interpersonal relations on crew performance during space simulation studies. *Life Support and Biosphere Sciences, 5*(4), 461–470.

Sandal, G. M. (2004). Culture and tension during an international Space Station simulation: Results from SFINCSS'99. *Aviation, Space, and Environmental Medicine, 75*(7), C44–C51.

Sandal, G. M., Vaernes, R., & Ursin, H. (1995). Interpersonal relations during simulated space missions. *Aviation, Space, and Environmental Medicine, 66*, 617–624.

Sandal, G. M., Vaernes, R., Bergan, P. T., Warncke, M., & Ursin, H. (1996). Psychological reactions during polar expeditions and isolation in hyperbaric chambers. *Aviation, Space, and Environmental Medicine, 67*(3), 227–234.

Santy, P. A. (1994). *Choosing the right stuff: The psychological selection of astronauts and cosmonauts*. Connecticut: Praeger.

Santy, P. A., Holland, A. W., Looper, L., & Marcondes-North, R. (1993). Multicultural factors in the space environment: Results of an international shuttle crew debrief. *Aviation, Space and Environmental Medicine, 64*(3), 196–200.

Stuster, J. (1989). Habitability during long-duration space missions – Key issues associated with a mission to Mars. In *Proceedings from American Astronomical Society Conference*, Boston, MA, January 1989.

Stuster, J. (1996). *Bold endeavors: Lessons from polar and space exploration.* Annapolis, MD: Naval Institute Press.

Stuster, J. W. (2010). Behavioral issues associated with isolation and confinement: Review and analysis of astronaut journals. *National Aeronautics and Space Administration. NASA/TM-2010-216130.*

Stuster, J., Bachelard, C., & Suedfeld, P. (2000). The relative importance of behavioral issues during long-duration ICE missions. *Aviation, Space, and Environmental Medicine, 71*, A17–A25.

Suedfeld, P. (2003). Canadian space psychology: The future may almost be here. Canadian Psychology, 44, 85–92.

Vanhove A. J., Herian, M. N., Harris, P. D., Luthans, F., & DeSimone J. A. (2014). *Psychosocial well-being and performance: Final report* (White paper). NASA/TM-2015-218565 prepared for NASA Johnson Space Center, Houston, TX: Behavioral Health and Performance. Retrieved May 18, 2020, from https://www.researchgate.net/publication/274254589_Examining_Psychosocial_Well-being_and_Performance_in_Isolated_Confined_and_Extreme_Environments

Vessel, E. A., & Russo, S. (2015). *Effects of reduced sensory stimulation and assessment of countermeasures for sensory stimulation augmentation: A report for NASA behavioral health and performance research: Sensory stimulation augmentation tools for long duration spaceflight* (NASA/TM-2015-218576). NASA-Johnson Space Center, Houston, TX. Retrieved May 18, 2020, from http://kinetic-cna.pl/web/pageFiles/kcfinder/files/adaptacja%20synaptyczna%20ARc.pdf

Weitz, P. J. (2000). *Oral history transcript – Interview with Paul J. Weitz.* [interv.] Rebecca Wright. 26 March 2000.

Whitmore, M., McGuire, K., Margerum, S., Thompson, S., Allen, C., Bowen, C., Adelstein, B., Schuh, S., Byrne, V., & Wong, D. (2013). *Evidence report: Risk of incompatible vehicle/habitat design.* Houston, Texas: NASA Human Research Program Space Human Factors and Habitability Element, National Aeronautics and Space Administration, Lyndon B. Johnson Space Center.

Winisdoerffer, F., & Soulez-Larivière, C. (1992). Habitability constraints/objectives for a Mars manned mission: Internal architecture considerations. *Advances in Space Research, 12*(1), 315–320.

Chapter 4
Review: Studies and Architecture of Habitability Missions in Mockups and Simulated Environments

This chapter and Chap. 5 introduce and outline important milestones of the journey that lead to increasing the fidelity of habitability studies intended to eventually allow us to design the best-fit extraterrestrial habitats. Beginning with the utilization of mockups and simulated environments in this chapter and in-situ environments in Chap. 5, an overview of exemplar habitats and their associated research that have contributed to studies of extraterrestrial habitability and human factors is presented. For each exemplar habitat, the physical and social settings, as well as mission relevant details, are summarized. Images and schematic plans illustrate the facility layout (see also Fig. 1.2). For readability and comparability reasons plans have been color coded according to primary human activities. Both short and long duration contributions to knowledge are briefly discussed to emphasize how habitability has slowly emerged as a central focus in the research on human adaptation and wellbeing in extraterrestrial living spaces. The chapter is structured in alignment with the current NASA Technological Readiness Levels. Short guest contributions by astronauts and simulation crewmembers complement the information.

4.1 Increasing Fidelity for Habitability Studies

Nothing illustrates better than concrete examples. The prior chapters have described the evolution of design considerations for extraterrestrial habitability as first primarily centered on surviving the place (particular environment) to provisions of critical elements to support thriving in the optimal living space. Organized in terms of fidelity levels defined by NASA Technological Readiness Levels (TRLs), the underlying driver for the transition from place to space was the necessity to transition from short duration missions to long duration missions. Table 4.1 provides a visual overview of this chapter and Chap. 5 and at the same time a quick overview of facilities with relevant habitability studies and missions in relation to the NASA TRLs.

S. Häuplik-Meusburger, S. Bishop, *Space Habitats and Habitability*, Space and Society, https://doi.org/10.1007/978-3-030-69740-2_4

Table 4.1 Overview of habitability studies and relevant missions in relation to NASA technological readiness levels

Habitability studies/ Relevant missions				
	Early missions	Laboratory environment	In-Situ relevant environment	Space environment
TRL level		1–4	5–6	7–9
Chapter	4.2 Early Missions	4.3 Habitability Missions in Mockups and Simulated Environments	5.1 Habitability Studies in-situ Terrestrial Environments	5.2 Habitability Studies in Extra-terrestrial environments
		Chambers, field locations		
Short term <30 days		Biodome missions, HERA, IBMP	Tektite I & II, HMP, FMARS, MDRS, HI-SEAS, Aquarius/ NEEMO	Salyut, Spacelab module, Skylab, Mir, ISS
Long term	Fram	Biosphere 2, IBMP	Concordia, HI-SEAS 1–5, FMARS, South Pole, Vostok	Salyut, Mir, ISS

Not only did the shift from short duration missions to long duration missions require new facilities to be built to meet long duration mission demands, it also forced existing facilities to be reorganized in terms of capability and fidelity for the purposes of research. The increasing fidelity of habitability research is intimately tied to these expanded capabilities of various habitats. The resulting broad spectrum of habitats utilized for both opportunistic ad hoc research as well as controlled studies necessitated an additional organizing schema that specifically located different habitats on a continuum of research fidelity. Cohen (2012) outlined standard research guidelines for Space Habitat Analogues to help systematize design and research efforts over this broad spectrum of habitability types (see Fig. 4.1).

Thus, early facilities have contributed to both short and long duration habitability knowledge as they evolved in response to the importance of critical mission characteristics. The following two chapters, therefore, are intended to highlight some of the best examples of various types of analogue and extraterrestrial habitats in order to characterize what is known as well as what is unknown to be addressed in subsequent chapters. Starting with an early example of an isolated and confined historical mission (Sect. 4.2), examples of missions in mockups and simulated habitats are covered next (Sect. 4.3). In Chap. 5, facilities situated in-situ environments, i.e., real environments posing some environmental challenge and isolation (Sect. 5.1) and those extraterrestrial facilities that have encompassed habitability foci to date will be addressed (Sect. 5.2).

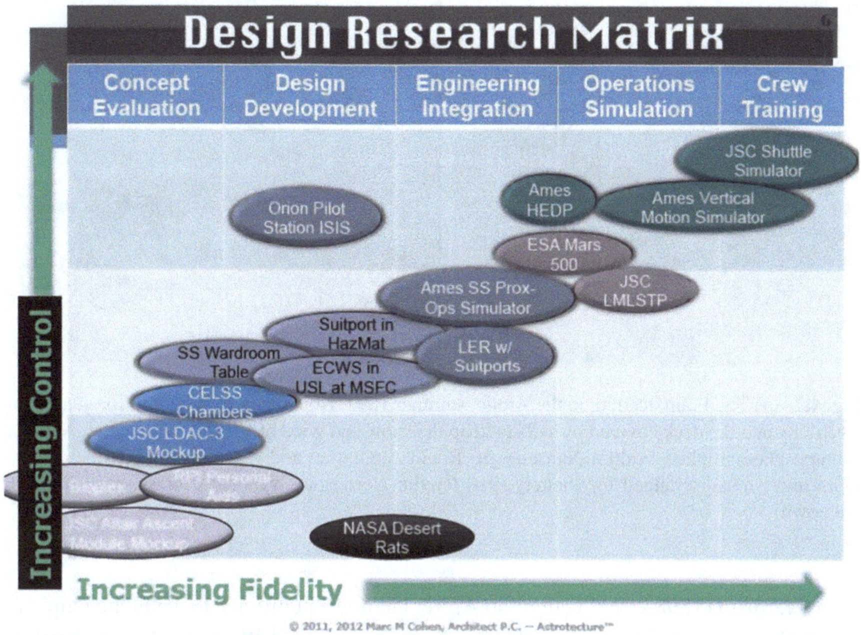

Fig. 4.1 Cohen's Design Research Matrix for assessing control and realism across the five purposes for mockups and simulators (credit: Marc M. Cohen)

4.2 Early Missions: Arctic Expeditions with Fram

> Never stop because you are afraid—you are never so likely to be wrong. Never keep a line of retreat: it is a wretched invention. The difficult is what takes a little time; the impossible is what takes a little longer. Fridtjof Nansen (Cool Antarctica 2001)

One approach in recent times has been to revisit the early exploratory missions applying qualitative analyses of the written records examining crew dynamics, habitability factors, psychological functioning under extreme conditions and adaptation to isolation, confinement or prolonged survival situations (Stuster 1996; Dudley-Rowley et al. 2001). These missions are difficult to compare, because they varied widely in duration from short excursions covering a few weeks to important (very) long duration missions, and the primary mandate for these initiatives was exploration with no original intention of conducting research on the crew themselves. However, the crew narratives anecdotally reveal many instances in which habitability played essential roles in both functional and dysfunctional adaptation to the extreme and often challenged circumstances in which the crew found themselves.

The ship Fram (see Fig. 4.2a, b) was an expedition ship built by Colin Archer and used by Norwegian explorers Fridtjof Nansen (1893–1896 Arctic Expedition) and

a b

Fig. 4.2 (**a**) The Fram, iced in, in the Arctic, summer 1894. Note the large windmill for generating electricity and the deck covered by sails to keep the snow and wind off (credit: Norwegian National Library, photographer: Fridtjof Nansen). (**b**) Roald Amundsen and Martin Ronne on the deck of Fram under a sail arranged for shelter, 1910 (credit: Norwegian National Library, photographer: unknown)

Otto Sverdrup (1898–1902 Canadian Arctic Island Expedition). In 1909 the ship was repaired and adapted to survive the forces of the Arctic drift-ice for Oscar Wistling and Roald Amundsen's expedition (1910–1912, South Pole Expedition) (Cool Antarctica 2001).

4.2.1 Fram Architecture and Design

The Fram was a 3-masted fore-and-aft schooner with an overall length of 39 m (128 ft), a width of 11 m (36 ft) and a weight of 420 tons (see Fig. 4.3). It had a 180 horse-power Diesel engine which was calculated to give her a speed of 4.5 knots. The ship was built of American elm and oak timbers, lined with pitch-pine. Deck planks were made of Norwegian fir. The wooden construction was strengthened with galvanized iron plates and the sides were stiffened with diagonal stays to withstand the ice-pressure.

The Fram was built to house a crew of 13 with provisions for up to 6 years and fuel oil for 8. Cabins and workrooms were insulated with wooden paneling, the interstices filled with cork, reindeer hair, thick felt and linoleum. With time, damp that penetrated the material produced fungus and decay of the wood. Petroleum was used to warm the vessel. Electric power for lighting was provided by a windmill once they were ice-bound.

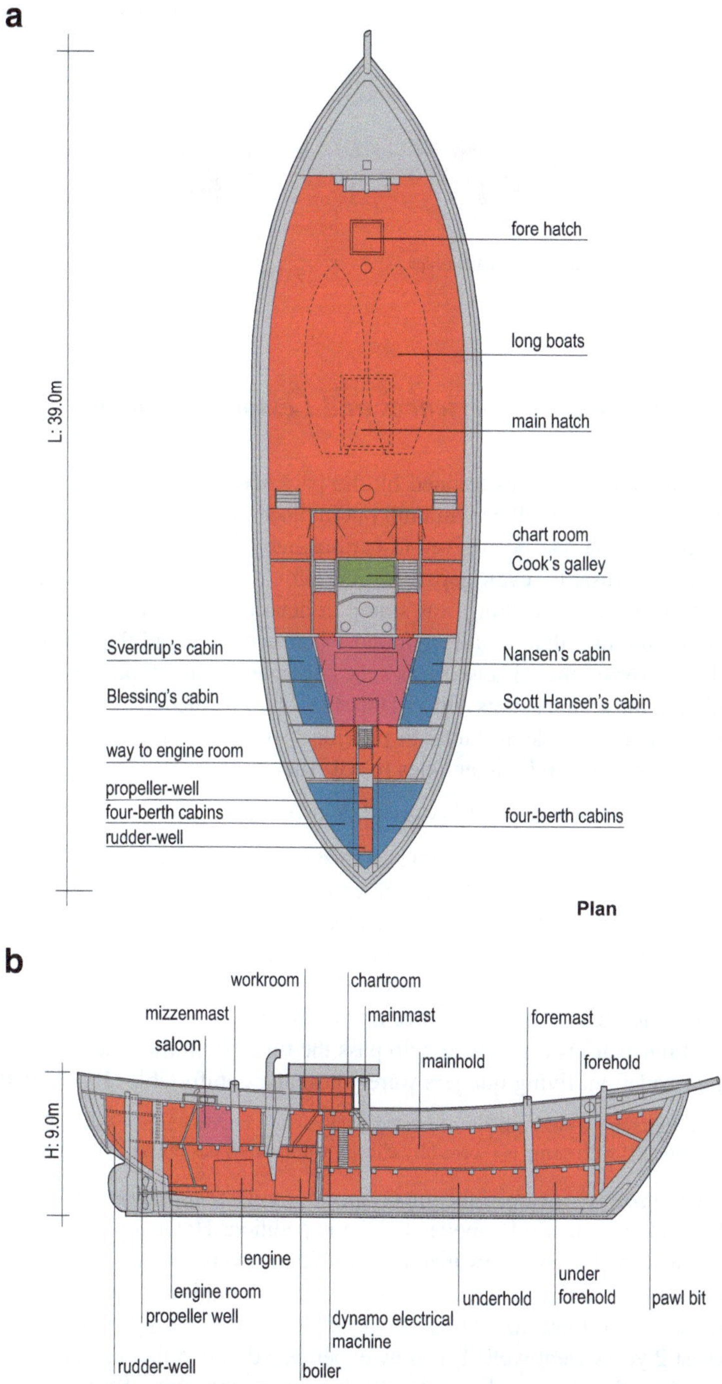

Fig. 4.3 (**a–d**) Schematic plan and section of the FRAM ship (credit: The Authors)

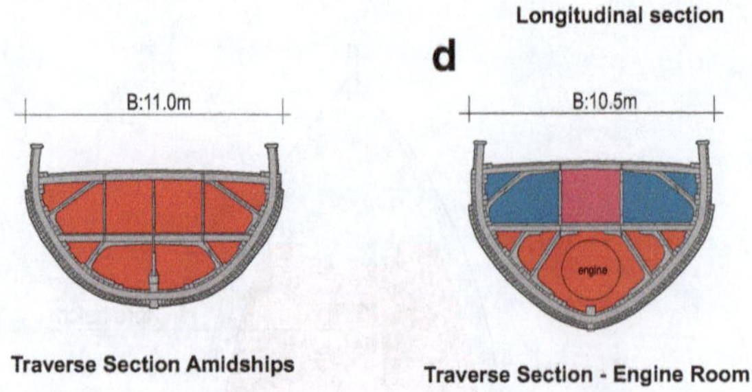

Fig. 4.3 (continued)

4.2.2 Habitability, Experiences and Lessons Learned

In 1893, once the Fram was trapped by the pack ice, the rudder and engine were dismantled and stored, and the windmill put in place to provide electric power to the ship. Great attention had been given to various habitability factors. Every available space was made usable, even gaps between barrels were filled with firewood. To keep track of where everything was stored, a detailed manifest of the entire cargo was kept so that any item could be found when needed. Applying his prior arctic expeditionary experience, Nansen was very cognizant of the impact of loss of color stimulation in the polar regions and had the ship's hull painted grey with bright green gunwales and red petroleum tanks to provide visual relief from the monochrome landscape of artic ice and winter skies (Cool Antarctica 2001).

> Ugh, I am tired of these endless white plains! (Nansen 2011, p. 426)

To counter "*ennui*" [French: boredom] Nansen enforced a fixed schedule of useful work which included meteorological research on the Aurora Borealis as well as systematic measurements on sea temperature, depth, and salinity, among others. There were sled dogs to take care of and tend which also provided companionship. Deep attachments were formed with the dogs which created unavoidable distress when it was later necessary to sacrifice them. There was a well-stocked library with games and musical instruments to help pass the time. With the extensive reinforcement and insulation, living quarters were warm and comfortable (Ryne 2020).

> Below decks were a cozy saloon decorated with carved dragon's heads and a library that Nansen stocked with some 600 carefully chosen volumes. (Sides 2009)

Nansen was also very aware of the importance of food and took great care in selecting the right food and plentiful supplies for the expedition. However, he did not hire a professional cook. Instead, crew members took turns every 4 weeks which shared the burden of cooking and gave each crew member opportunities to contribute. "*Juell, at least, was a find*" (Huntford 1997, p. 2) as he had learned baking from his wife.

The first 2 years went well. The crew experienced only a few mishaps, including two dogs lost due to a polar bear attack outside the ship. Still, isolation and confinement took a heavy toll on the crew. Diary entries clearly showed evidence of individual distress with isolation but also of inevitable intracrew conflict that comes when individuals are confined in close quarters for extended periods of time. Amundsen once described the ship Fram, as "*this cold prison of loneliness in which we are confined*" (Huntford 1997, p. 34).

Nansen was a strict opponent to drinking alcohol on board, which contributed to feelings of personal distress among the crew (Kobak 1999). The ship's doctor, Henrik Blessing, secretly became addicted to morphine trying to cope. The tension was succinctly summarized by Nansen himself as "*Nerves taut from the time one wakes up in the morning until one closes one's eyes late at night*" (Kobak 1999). A ubiquitous theme throughout the entire crew's diaries, including Nansen's, was their expressions of missing family and friends.

Eventually, the slow drift progression of the ship, tension between crew members and boredom moved Nansen to change his plans to a two-man dash for the North Pole (Nansen and Johansen). Although they were now free of the confines of the ship, the overland attempt to reach the North Pole was fraught with difficulties and challenges. They spent 3 weeks trying to get to the North Pole and 14 months getting back to safety. The necessity to hunt walruses, polar bears, seals and foxes for meat, blubber, bones and fur without an adequate base to operate from created serious conditions of filth from grease and soot from cooking with blubber, clothes that were frequently frozen or mushy, sharing a sleeping bag of rotting reindeer fur and incidences of great risk dealing with immersion in freezing water and losing their bearings. Their diaries reflect the realization that Nansen and Johansen found they had nothing in common and often traveled in silence further exacerbating feelings of isolation and loneliness. Ironically, although they broke the 400-year-old record for the furthest North by 146 miles, they were still 230 miles short of the pole when they decided to turn south for home. The Fram, on the other hand, continued to drift for another year and a half before breaking free of the pack ice and covered two-thirds of the distance Nansen and Johansen covered on foot. In the end, the two men got only 19 miles farther north than they would have if they had stayed on the drifting Fram (Kobak 1999; Cool Antarctica 2001).

The various expeditions aboard the Fram exemplify shipboard challenges in general. Close quarters, limited space and resources, and an extremely volatile environment are the hallmark characteristics of all vessels. The Nansen expedition was different in that the period of being ice-bound provided an expanded space. This additional operational space decompressed the space limitations to some extent. Once evidence that the sea ice would not crush the ship, the ship became a stable safe-haven subject only to the environmental weather conditions rather than both sea and weather. Many other sea-going research vessels remain subject to the physical challenges posed by sea conditions. Ongoing efforts to address habitability factors in more modern times have primarily advanced in the areas of communication allowing crew members to remain in touch with family members and interact with the larger world community. Such connectivity remains a key countermeasure to feelings of isolation across all ICEs.

4.3 Habitability Missions in MOCKUPS and SIMULATED Environments

To conduct true research[1] requires control over the environment in which crew members operate. Experimental research on habitability factors for ICEs was initially conducted in highly controlled habitats (i.e., laboratory or capsule research) that simulate various components and did not require or involve a complete architectural setting. Often only parts of a habitat were simulated and tested, e.g., the work area or laboratory facilities.

With reference to the Technology Readiness Level (TRL), the classification levels used at NASA (NASA [TRL] 2017), this group of studies encompass the TRL Levels 1–4:

- TRL 1—Basic principles observed and reported
- TRL 2—Technology concept and/or application formulated
- TRL 3—Analytical and experimental critical function and/or characteristic proof-of-concept
- TRL 4—Component and/or breadboard[2] validation in laboratory environment

With the demand for longer duration missions and the transition from *place* to *space* (Chap. 2), the necessity for development of more extensive and complete facilities was recognized and met. The TRL system alone does not adequately encompass the defining characteristics that cover the spectrum of mockups and simulation facilities. As shown in Fig. 4.1, simulated missions can run the gamut of pure laboratory setups involving one or two rooms located within the safe confines of an institutional setting with full safeguards in place and typically used for short duration missions (Sect. 4.3) to more complete facilities located in actual challenging environments (in-situ) with the attendant higher risk and supporting longer duration missions (Chap. 5).

Nevertheless, they share a common research focus as to their primary goal and are designed to provide a higher degree of control over extraneous factors than opportunistic ad hoc research conducted in largely operational environments such as existing polar bases in Antarctica (e.g., McMurdo, see NSF 2015). The defining characteristic for the mockups and simulation group addressed in this chapter is that they are situated in an institutional setting (e.g., university, research institute or space agency center) in which *full control is exercised over all extraneous factors and full*

[1]Various definitions for 'true research' exist and involve the experimental framework. In general, a hypothesis is tested for evidence to support the hypothesis or fails to find supportive evidence. True research includes only a few (preferably, only one) variables that can be manipulated and therefore tested, with randomly assigned sample groups (including a viable control group).

[2]Breadboard validation: A low fidelity system/component **experimental model** is built and operated to demonstrate basic functionality and critical test environments, and associated performance predictions are **defined** relative to the final operating environment. Laboratory environments refer to any facility specifically designed to provide controlled conditions in which scientific or technological research, experiments, and measurement may be performed.

safety guarantees are in place, eliminating both environmental risk as well as true isolation. In essence, research conducted in these facilities would be characterized as confinement studies.

4.3.1 Biodome Missions: Closed Ecosystems

Closed ecosystems share common atmospheric, water- and waste recycling systems. Usually the design (and operation) of a closed ecosystem facility requires an extensive degree of science and discipline coordination spanning technology with ecology, biology and human factors. In the following sections, an early soviet/Russian facility (BIOS-3) as well as a facility in the US (Biosphere 2) are described as exemplars.

4.3.1.1 Bios-3

BIOS-3 is a closed ecosystem at the Institute of Biophysics in Krasnoyarsk, Russia "*that was designed to include humans in a functioning, self-sustaining, closed ecosystem for continuous periods as long as six months*" (Salisbury et al. 1997, p. 575) (see Fig. 4.4a, b). Its construction began in 1965 and was completed in 1972, becoming the first experimental ecosystem that included humans as the primary test subjects rather than simply a source of metabolic data (Nelson et al. 2010, p. 31). The BIOS-3 facility was used from December 1972 to 1984. In 1991, it became part of the International Center for Closed Ecosystems, a subdivision of the Institute of Biophysics, Siberian Branch but remained largely inactive. In 2005 closed

a **b**

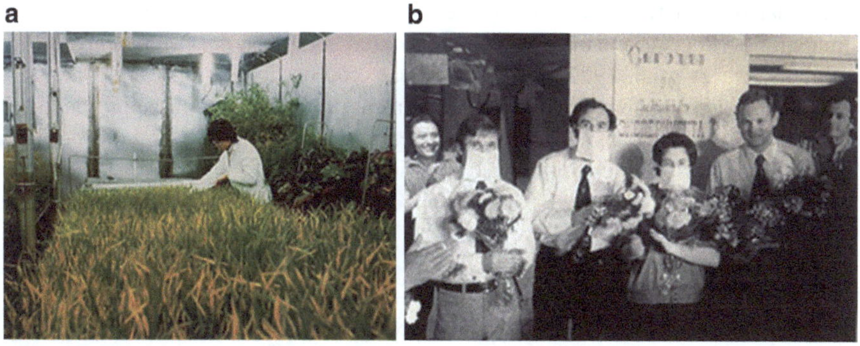

Fig. 4.4 (**a**) Inside the BIOS-3 facility (credit: IBP, Institute of Biophysics, Russian Academy of Sciences, Siberian Branch). (**b**) Three test subjects (wearing respiratory masks) and doctors meeting them when they leave Bios-3 after the 180-day experiment (credit: IBP, Institute of Biophysics, Russian Academy of Sciences, Siberian Branch)

ecosystems research focusing on growing plants and recycling waste was restarted, in cooperation with the European Space Agency.

Facility Architecture and Design
BIOS-3 is suitable for up to three people and consists of a 315 m^3 (11,124 ft^3) underground stainless-steel structure to provide a hermetic seal. It is accessible from the main building of the Institute of Biophysics by a passageway. The 14.0 × 9 × 2.5 m structure (see Fig. 4.5) is divided into four airtight compartments (each about 79 m^3), each having tightly sealed doors. Each compartment can be sealed independently. The crew compartment provides three separate sleeping rooms, a kitchen, a lavatory, a control room, and work and recreational space (Salisbury et al. 1997). The toilet area serves also as an entrance airlock.

For the first experiments, two compartments consisted of two phytotrons (laboratories for growing plants under strictly controlled conditions) for growing wheat and a selection of vegetables; one compartment was an algae cultivator, and the other one the living compartment for the crew.

Habitability Studies, Experiences and Lessons Learned
Between December 1972 to 1984, three closure experiments were conducted with a two to three-person crews. The first experiment involved two men and one woman and lasted 180 days, during winter 1972–1973. The second experiment with three male crew members (winter 1976–1977) lasted 4 months. The third experiment with two male crew members lasted 5 months (winter 1983–1984). Scientist Yevgeny Shepelev became the first test subject himself and, thus, the first human to survive 24 h in a closed ecological system. The results of this experiment were used to refine the simulator to produce and reuse almost 95% of the water needed by three people (Science Illustrated 2012). Condensate water collected from the plants was purified and provided as drinking and sanitary water. The sanitary water and liquid wastes from the crew were processed by the photo reactors (Nelson et al. 2010).

The harvested wheat grains were used for baking bread. Vegetables produced provided fresh food for the crew. In addition, freeze-dried food as a necessary source of animal protein was stocked. The researchers included animal products in all simulations. Vegetarianism was not considered as "*Siberians must have their meat*" (Salisbury et al. 1997, p. 581). Because crop production efficiency was low (10%), the scientists concluded that enough meat protein needed to be provided for Lunar and Mars missions. Overall, crew members stayed healthy and did not lose weight (in contrast to the Biosphere 2 experiments).

4.3.1.2 Biosphere 2

Biosphere 2 was an Earth system science research and development facility "*designed to study global ecology and to test bioregeneative life support on a biospheric scale*" (Nelson et al. 2010, p. 530). The facility was constructed between 1985 and 1991 in Arizona, USA, and used for two closure experiments from 1991 to 1994 (see Fig. 4.6a, b). Biosphere 2 was originally meant to be operated on a long-

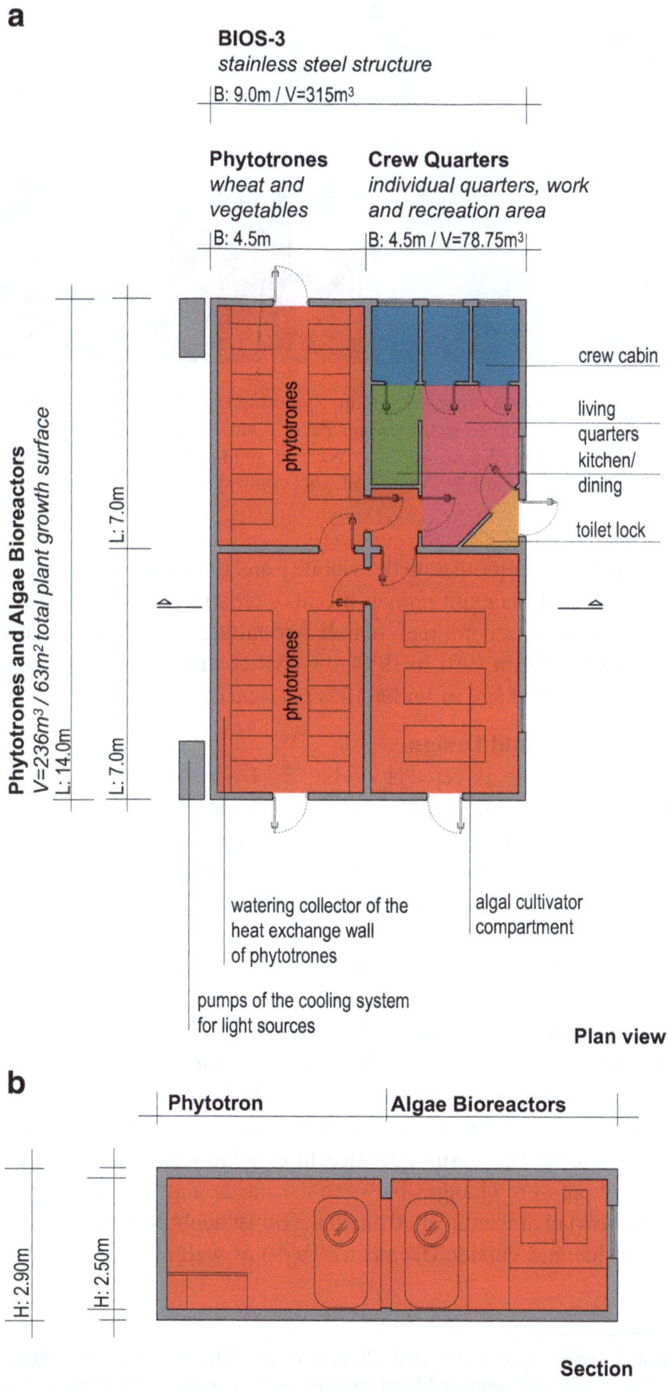

Fig. 4.5 (**a–b**) Schematic plan and section of BIOS-3 (credit: The Authors)

Fig. 4.6 (**a**) A rainbow over Biosphere 2 (credit: Spaceship Earth by Matt Wolf, shot courtesy of NEON). (**b**) Biospherians (left to right): Bernd Zabel, Linda Leigh, Taber MacMullen, Abigail Alling, Mark Van Thillo, Sally Silverstone, Roy Walford, and Jane Poynter posing inside Biosphere 2 in 1990 (credit: Spaceship Earth by Matt Wolf, shot courtesy of NEON)

term basis to study and demonstrate the viability and dynamics of a closed ecological system to support and maintain human life in outer space. The system consisted of seven biomes, including a coral reef, marsh, agriculture, rainforest, savannah, desert and a human habitat. From 1991 to 1994, the facility was used as a materially closed system (with a small leakage of under 10% per year).[3]

Facility Architecture and Design

The Biosphere 2 facility covered 1.27 ha (3.14 acres) and had an atmospheric volume of over 180,000 m^3 (6,356,640 ft^3) (Dempster 1999). Its seven biome areas included a 1900 m^2 (20,000 ft^2) tropical rainforest, an 850 m^2 (9100 ft^2) ocean with a coral reef, a 450 m^2 (4800 ft^2) mangrove wetlands, a 1300 m^2 (14,000 ft^2) savannah grassland, a 1400 m^2 (15,000 ft^2) fog desert, and two anthropogenic biomes: a 2500 m^2 (27,000 ft^2) agricultural system and a human habitat with living spaces, laboratories and workshops (see Fig. 4.7).

The architectural skin was made of a steel spaceframe and double laminated glass panels that materially isolated the facility. The so-called Technosphere was underground and housed all electrical plumbing, and mechanical systems. It was materially isolated by steel sheets. To study the ecological processes inside a closed ecological system, the degree of closure (with minimum atmospheric leakage) had to be high. In an airtight enclosure, the structure had to withstand fluctuations due to temperature and pressure changes between the inside and the outside. That is why the facility has two chambers, called 'lungs'. Those neutralize the pressure changes between the inside and outside (Dempster 1999) as well as allowing the heated air

[3]With an annual leakage rate under 10% (Nelson et al. 2010, p. 530). More information and publications from the experiments and from the operation can be found at https://ecotechnics.edu/publications/ and https://biosphere2.org.

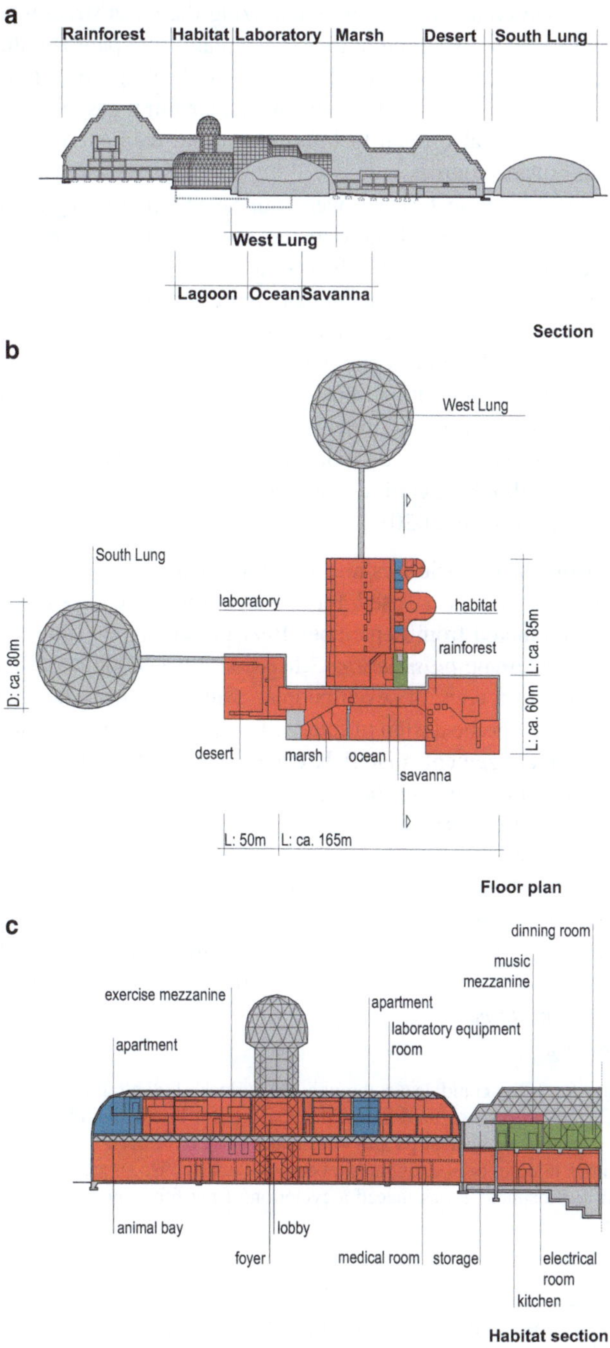

Fig. 4.7 **(a–c)** Schematic plan and section of BIOSPHERE 2 (credit: The Authors)

inside to expand and contract without jeopardizing the glass structure. The facility was energetically open to sunlight to enable plant life photosynthesis without artificial light and to convert thermal energy for heating. Additional electrical power was supplied into Biosphere 2 from an onsite natural gas energy center (UoA 2020). The facility was also open to information and data exchange between the crew inside and outside sources.

After Mission 1 technical issues were improved. Interestingly the decline in atmospheric oxygen was traced back to the use of untreated concrete and soils which absorbed the oxygen (Severinghaus et al. 1994).

Special attention was given to the human quarters with windows that looked out over all other parts of the Biosphere. There was laboratory space, a medical clinic, exercise and recreational facilities and a communications center. Every crew member had a private apartment with an upstairs bedroom and downstairs living room. Their bathrooms had showers and toilets although shower time was limited. Drinking water was condensed recovered moisture produced by the plants in Biosphere. 'Household chores' (kitchen, cooking and cleanup duties) were shared by all crew members (Science.jrank.org 2020).

Habitability Studies, Experiences and Lessons Learned

The facility Biosphere 2 was used for 4 years for closed-system experiments. Mission 1 was conducted from September 1991 to September 1993 with a crew of four male and four female *biospherians.*[4] Mission 2 was conducted from March to September 1994 with a crew of seven people.[5] Much has been learned from these two closure missions for the design of future life support systems, particularly for the water and waste management, materials cycles and food production. "*More than 3000 species, excluding soil microbiota [...] were selected over a period of 5 years of research and development*" (Eckart 1996, p. 367).

Major advances were made concerning habitability and human factors in an isolated and confined environment from the Biosphere missions. These isolation missions set a new record. "*No one knew if we could stay inside for two years; so many things could go wrong*" (Nelson 2018a, b, viii). According to Mark Nelson, one of the biospherians of Mission 1, the experience inside the dome ranged from "*This facility is a nightmare*" to "*It's going to save the Earth and colonize space*!" (Nelson 2018a, b).

> At any time, [...] we could leave through the airlock. I don't think any of us were tempted. We had worked so hard to get there, [...] But the strains of living in a small world with a limited number of people were real. [...] I have been to Heaven and Hell a myriad of times in here, been enlightened, been profoundly discouraged, you name it, I've experienced it. Biosphere 2 was indeed a cyclotron of the life sciences (Nelson 2018a, b, p. 199).

---

[4]The crew were: Roy Walford, Jane Poynter, Taber MacCallum, Mark Nelson, Sally Silverstone, Abigail Alling, Mark Van Thillo, and Linda Leigh.

[5]The crew was Norberto Alvarez-Romo (Capt.), John Druitt, Matt Finn, Pascale Maslin, Charlotte Godfrey, Rodrigo Romo and Tilak Mahato.

In both missions, tensions among the crew began early and led to the "Us versus Them" split common in isolated groups. Jane Poynter describes one of the coping strategies employed by Mission 1 crew member in Dynamics in Isolated Groups—Playing Theatre as a Coping Strategy (Box 4.1).

Box 4.1 Dynamics in Isolated Groups: Playing Theatre as a Coping Strategy

"On October 12 we celebrated Gaia's birthday with a rave in the command room. We all danced on the granite conference table, and performed strip shows, though as far as I recall, not one of us stripped down to the full monty! Our spirits had been hoisted by some fluorescent green ethyl alcohol Roy magically produced from the medical facility. A friend had smuggled it past inspection at a recent import/export by dropping food colouring in the alcohol and labeling it "preservative."

People have often asked how we kept going despite the upheavals. I think it was all the theater training that allowed us to behave so differently in these various settings. Tony Burgess recalls a conversation with Mark Nelson years ago at the Caravan of Dreams. "Mark said, 'We do whatever we need to do to get done.' That was the clue I got to the dynamics of your group. It was all theater. Play a part and play it well."

Source: Poynter 2006, p. 237

4.3.2 *The IBMP Facility*

The Moscow Institute of Biomedical Problems (IBMP) has had an isolation facility in operation since 1967 which has conducted dozens of long duration studies beginning with a year-long study for its first mission. The most famous to date was the Mars 500, a collaborative project between Russian Space Agency (RSA) and European Space Agency (ESA) starting in 2007 with completion of the 520-day mission in 2011 (see Fig. 4.8a, b). Later in 2015, an all-female crew mission that

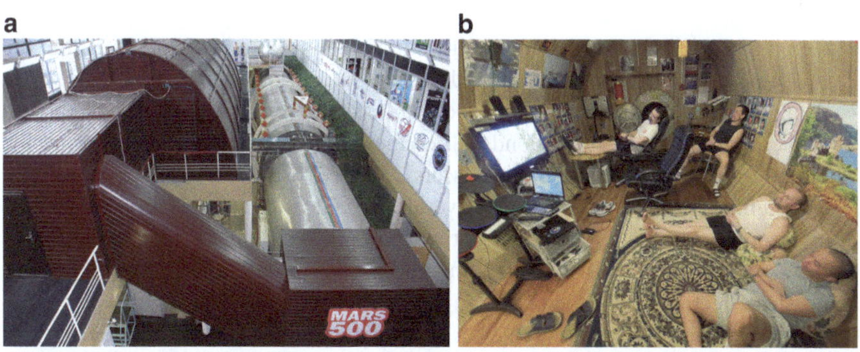

Fig. 4.8 (**a**) Image of the isolation facility at IMBP during the Mars 500 mission (credit: ESA—S. Corvaja). (**b**) Four crewmembers are relaxing in the Mars500 Habitat module (credit: ESA)

lasted 8 days took place and longer missions for 4, 8 and 12 months have been planned (Orlov et al. 2015).

Facility Architecture and Design

For the Mars 500 experiment, the isolation facility (see Fig. 4.9) was comprised of four hermetically sealed interconnected habitat modules: the Medical Module (EC-100), the Habitat Module (EC-150), the Technical Module (or Utility and storage module, EC-250) and the Mars landing simulator (EC-50). In addition, one external module was used to simulate the 'Martian surface'. The total volume of the four modules was 550 m^3 (19,423^3 ft) (ESA [Mars500] 2012). The Habitat module hosted individual compartments for the 6 crewmembers, a social space, kitchen and lavatory. The Utility module provided a gym, greenhouse, storage for resources, a fridge, and another lavatory. The Medical module was equipped with a habitable compartment, a kitchen-dining room, as well as working places with medical equipment and a lavatory.

Habitability Studies, Experiences and Lessons Learned

The longest duration experiment conducted at the IMBP facility was the Mars 500, a 520-day isolation experiment that took place from April 2010 to October 2011. The Mars 500 experiment has been the longest terrestrial simulation analogue mission to date in preparation for a human mission to Mars and included a number of experiments of various durations. Pre-experiment missions included a 14-day isolation (completed in November 2007) and a 105-day isolation (completed in July 2009). The six person crew for the 520-day mission was comprised of three Russian members, two from the European Space Agency, and a Chinese participant, all male. The crew performed tasks planned according to a long-term mission timeframe. Considerations taken into account included labor intensity, complexity of required methodologies, crew members' specialties, and the possibility of experiments to cross-reference and influence each other, and observations whether some special conditions are present during an experiment's implementation (Haeuplik-Meusburger et al. 2017).

Some of the Mars 500 results from video analysis were summarized by Basner et al. (2015):

- Coping strategies to address monotony and boredom from low workload after the first mission quarter and to restricted social contacts are needed.
- A higher frequency of crew-perceived conflicts with mission control was reported in the first relative to the second half of the mission (being maximal during the period of the simulated landing on Mars).
- The number of crew interactions (overall amount of communication) with mission control and the number of negative and critical statements in crew messages increased during the simulated landing period. The researchers did not find a third quarter effect in any of the psychological or behavioral outcomes.
- Two crewmembers were notable for showing no signs of behavioral changes or psychological distress during the mission; they were most often mentioned as the two people with whom the rest of the crew interacted; and they were the only two

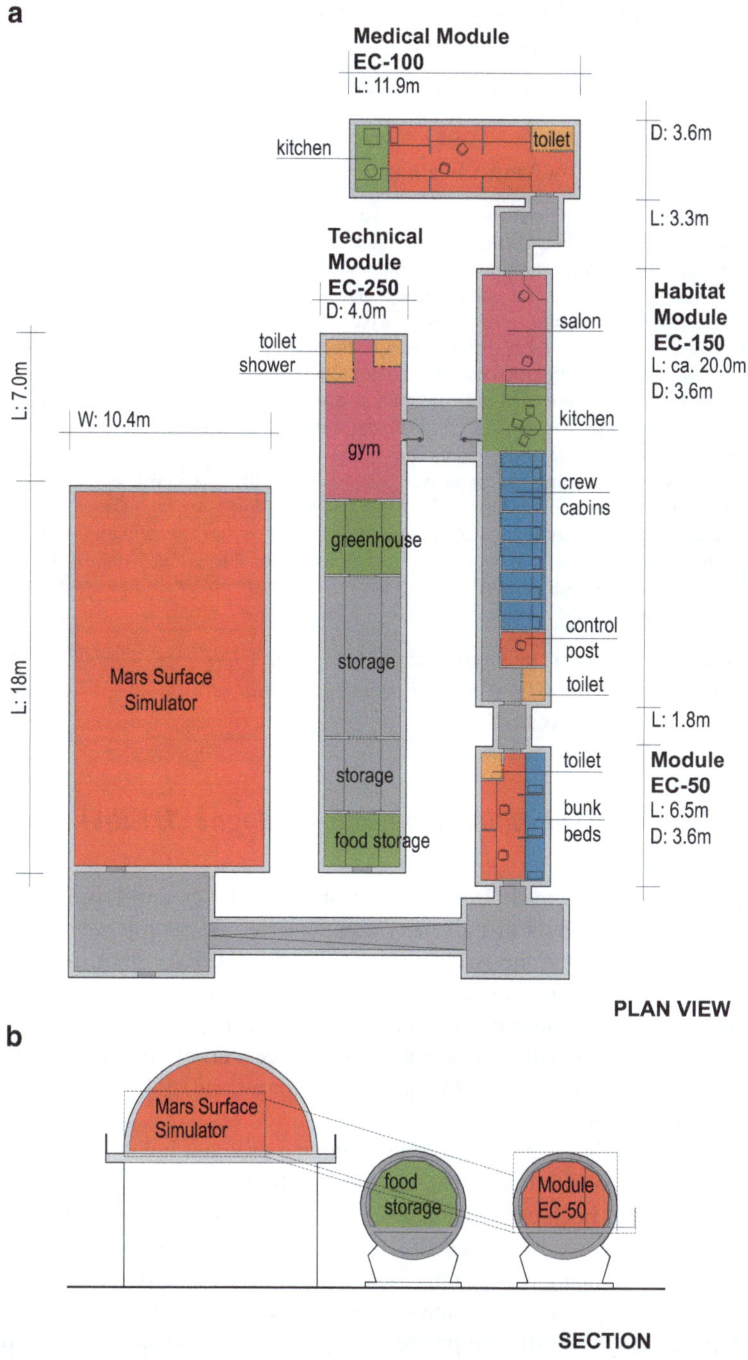

Fig. 4.9 (**a–b**) Schematic plan and section of the IBMP Isolation Facility (credit: The Authors)

crewmembers to suffer no changes in sleep duration, sleep-wake timing or sleep quality during the 520-day mission.

For a personal insight, see Box 4.2: Interview with Romain Charles.

Box 4.2 520 days on Mars: Feeling the distance

Definitely I never felt that I was in space. We had gravity, so I was on Earth. My mind knew that all the people around working on the mission were just 20 meters away, but very quickly, and especially when we didn't have direct contact, I started to feel a real distance. This feeling was quite strong and was there for the whole trip. [...] We talked about family, girlfriends or wives. There were no real taboos between us. We talked about a lot of different things. For most of the 520 days we didn't talk what we would do after. We were focusing on the present. We wouldn't look at the whole big picture. We tried to avoid that, you know, 520 days is a very long time. Inside the modules we are taking it on a day-to-day basis. We had a lot of activities, and on top of that, we all had additional personal activities to keep us busy. [...] It definitely was a strategy to be busy. Before I began the mission, I talked with a colleague, who was a submariner years ago. He said: 'First of all, be always busy'. Because when you don't have anything to do, you start to have some dark thoughts, it's becoming harder. And that is why I had many personal activities. And it worked. This mission was a success, because we were six when we entered, and we were still six, when we exit the modules.

Source: *From a personal interview with Romain Charles, Mars 500 crewmember in 2013. A radio broadcast on that interview can be listened to here:* https://cba.fro.at/271785

4.3.3 Human Exploration Research Analogue (HERA)

HERA was formerly named 'Deep Space Habitat' and was developed by NASA, then enhanced via a series of university competitions. As such it was used in the desert research studies in the Arizona desert (NASA [HERA] 2014, 2015). It is a research facility that is used as an analogue for the simulation of isolation, confinement, and remote conditions of mission exploration scenarios. Its goal is to address risks and gaps associated with human performance during spaceflight. Standard simulation durations are 14, 30, 45 and 60 days for a crew of 4, usually consisting of 2 men and 2 women.

A HERA campaign is defined as one integrated protocol involving a number of individual investigator studies that can be combined under a primary mission campaign scenario. Each campaign consists of four consecutive missions (i.e., four different crews), providing a total of 16 subjects for each study involved in the campaign. Early campaigns were only 7 days in duration but have grown in duration and now typically span at least 45 days to meet the demand for long duration studies (NASA [HERA] 2016). Mission protocols require all four subjects to participate in mission activities beginning with up to 16 days of pre-confinement activities (i.e., baseline data collection (BDC), training, informed consent); 45 days in-mission confinement activities (i.e., operational activities expected for an exploration mission along with research activities); and up to 7 days of post-confinement activities (i.e., post-mission data collection, debriefing). To generate conditions of isolation rather than pure confinement, test subjects have a number of restrictions of their activities and access to communications channels (i.e., no access to email, personal work, phone calls, or the internet; family conferences are scheduled weekly). News items are provided by Mission Control via periodic uplinks consistent with simulating space flight conditions. A mission control center (MCC) is staffed 24/7 to support HERA operations including surveillance, communications, computers and phones (NASA 2019) (see Fig. 4.10a, b).

a

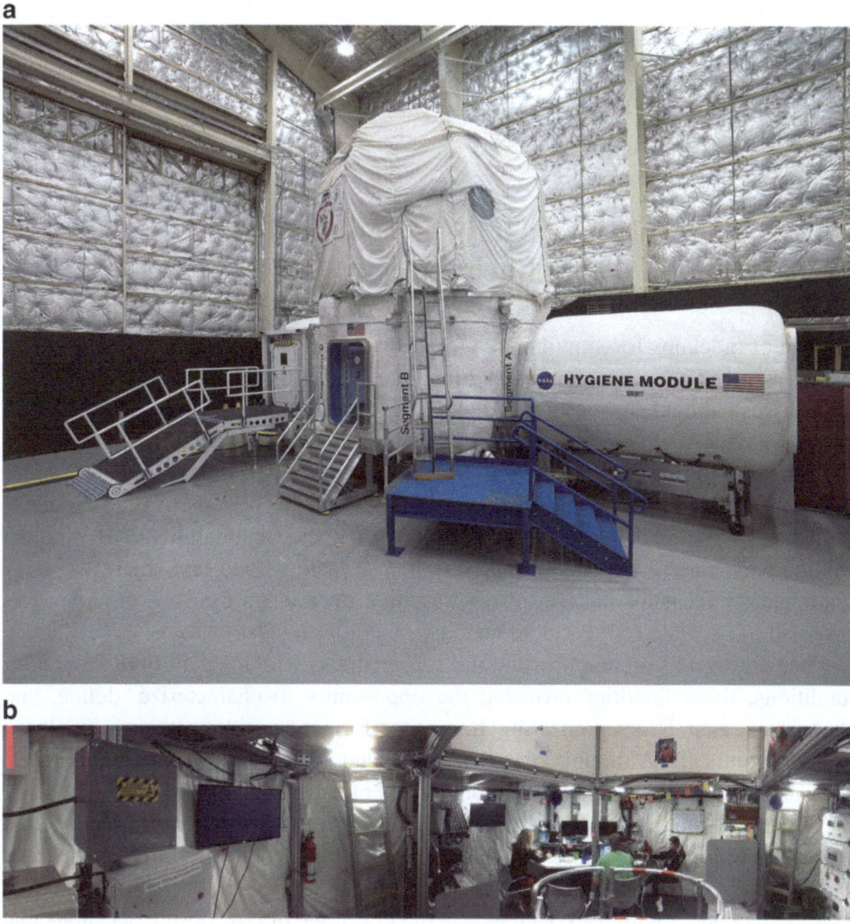

b

Fig. 4.10 (a) HERA—The Human Exploration Research Analog in the HRP (credit: NASA, photographer Bill Stafford and Robert Markowitz). (b) Inside HERA. The crew of HERA 12 (Mission 3 Crew 4) on the galley deck, celebrating the birthday of one of the crew members. The crew members are, left to right: Ulyana Horodyskyj, Todd Huhn, Jonna Ocampo. (credit: NASA, Mark Kerr)

Facility Architecture and Design

The Human Exploration Research Analog (HERA) is a three-story (two main stories and a loft), four-port habitat (see Fig. 4.11). The first floor or ground level comprises a medical science station, flight deck, maintenance workstation, operations console, and the attached hygiene module and airlock. The second floor houses a wardroom table, galley and aerobic exercise devices. This level is mostly for social and personal use. On the third loft level are the four crew quarters.

Habitability Studies, Experiences and Lessons Learned

There have been five HERA campaigns completed between 2014 and 2020 covering a wide variety of research foci encompassing around 100 studies on behavioral performance, team performance, crew medical and enabling technologies. As can be seen in Table 4.2, the focus has been heavily skewed towards team performance and enabling technology studies. The four habitability studies to date have primarily been development of enabling technologies rather than investigations of habitability impacts upon human functioning.

4.4 Summary of Habitability Missions in MOCKUPS and SIMULATED Environments

The use of mockups for simulations provided the early foundation of research into human centered adaptive response to confinement and modest isolation. Table 4.3 summarizes habitat features for the BIOS-3, Biosphere 2, Mars 500 isolation facility, as well as HERA.

Compared to prior laboratory and capsule environments, those simulation facilities were designed to accommodate humans over longer periods of time. Some of them have integrated particular provisions to enhance habitability. All of them provide personal crew quarters; the Biosphere 2 facility even had small two-story apartments. Except for the Biosphere 2 facility, even when the shower and toilets have had to be shared, often there were two facilities available.

Characterized by an emphasis on full scale habitats with the control of laboratory conditions, these facilities provided the opportunity to characterize, define, and identify relevant issues impacting human performance and wellbeing in closed loop environments. Although lacking true risk and isolation, they have been critical testbeds for technology and process development and validation.

With these facilities researchers and engineers have learned about incidents that no one could foresee. 'Failures' as much as successes provided rich lessons learned for the design of future habitats. It is impossible to anticipate all possibilities. The Biosphere incident with the concrete is perfect example of how design decisions that worked in the past have impacts in under new environmental requirements that can influence the outcome of the whole mission.

That is why it is important to build simulation habitats and use them with many different crews in order to provide the broadest opportunity to generate different

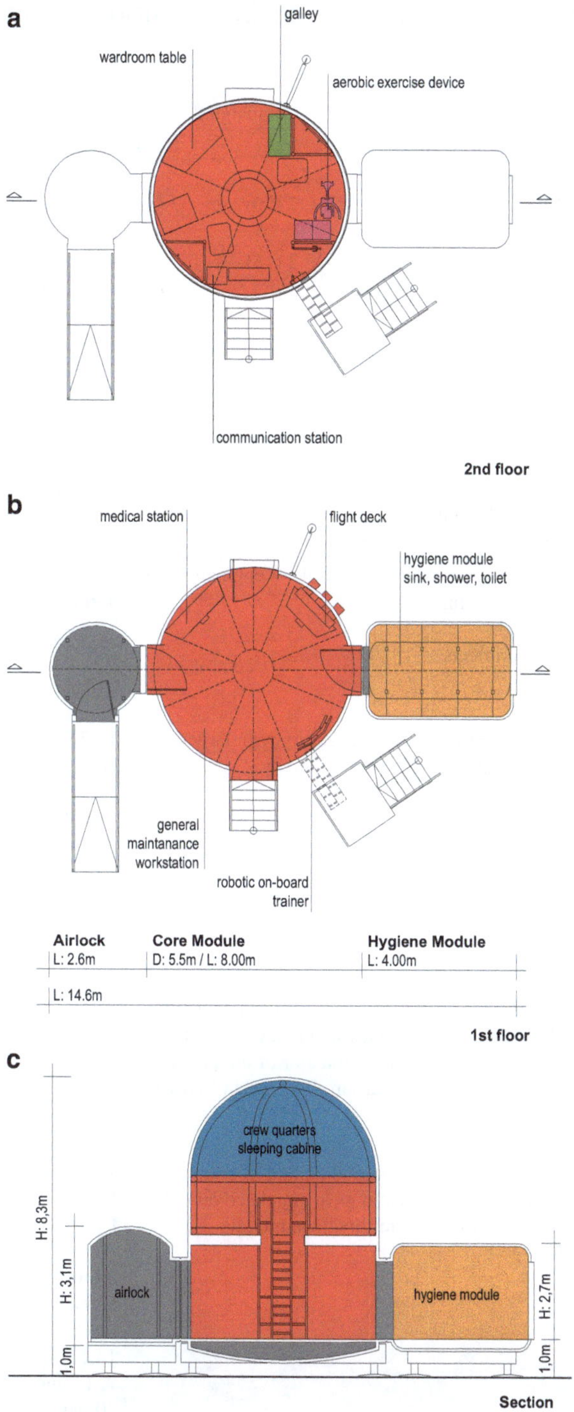

Fig. 4.11 (**a–c**) Schematic plans and section of HERA (credit: The Authors)

Table 4.2 Summary of research conducted at HERA 2014–2020

	Number of studies (%)					
Research Foci	Campaign 1	Campaign 2	Campaign 3	Campaign 4	Campaign 5	Campaign 6
Behavioral performance	0 (0)	1 (7)	10 (45)	3 (17)	2 (13)	1 (6)
Team performance	4 (31)	7 (50)	7 (32)	4 (22)	8 (53)	4 (25)
Crew medical	1 (8)	2 (14)	1 (5)	5 (28)	1 (7)	2 (13)
Enabling technologies	8 (61)	4 (29)	4 (18)	6 (33)	4 (27)	9 (56)
Total	13	14	22	18	15	16

Table 4.3 Summary of habitat features across mockups and simulated facilities

Facility/Module type	Spatial layout	Total area/ Simulation durations	Features
BIOS-3 Single structure; 4 airtight compartments	One module, 4 compartments	315 m^3 4–6 months	Long duration missions primarily intended to test an experimental closed loop ecosystem (enabling technologies) rather than human response/adaptation
BIOSPHERE 2 Large materially closed structure connecting 7 biomes	Large facility with largest spatial diversity	1.27 ha and atmospheric volume of 180,000 m^3 2 years, and 7 months	Long duration missions intended to test integrated closed loop ecosystems, agricultural production, dietetic health improvements, and the interaction of complex bionomic systems and atmospheric dynamics Largest variety in interior spaces and activity zones
IBMP Horizontal cylindrical, one-story	Various modules and activity zones within the whole facility; simulator of the landing Martian ship and landscape of Mars	550m^3 520 days	Psychosocial isolation experiment of a simulated human operated flight to Mars. Wood paneling of interior walls
HERA Vertical cylinder, two-story, four-port	Internal zoning of main activities; simulated airlock and hygiene module	148.1 m^3 Core: 56.0 m^3 Loft: 69.9 m^3 Airlock: 8.6 m^3 Hygiene module: 14.1 m^3 7, 14 and 30 days	Very short missions, limited suitability for human factors and habitability studies; Research on behavioral and team performance

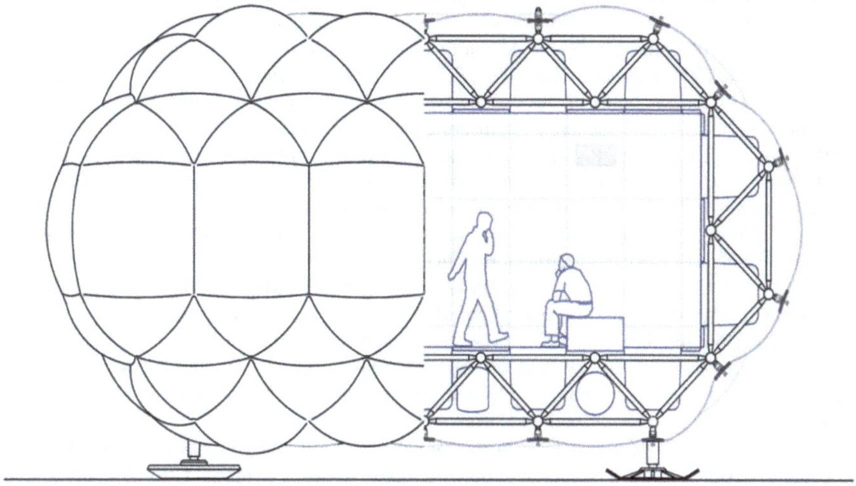

Fig. 4.12 Early Integrated Habitat design concept: Partial Transverse Section through the inflatable Integrated Habitat Structure. (Drawing credit: Andrew Daga, University of North Dakota)

results. Simulation facilities also have the advantage of testing different layouts and different spatial and social situations under conditions of some degree of safety and control. We not only need the possibility, but also the capability to thoroughly test the built structure.

This need is surprisingly unsatisfied at this point in time. Existing simulation facilities are largely static and not amenable to reconfiguration. The Lunar Daytime concept (Cohen et al. 2020) addresses the challenge behavioral scientists and architectural researchers face in conducting research in space habitats or habitat analogues to produce scientifically valid and reliable results. It proposes to build a sixth module for the already existing Multi-Purpose Research Station (MPRS) at the University of North Dakota (UND) as a relatively easy immediate intermediate solution to an analogue facility testbed (Fig. 4.12).

References

Basner, M., Savitt, A., Moore, T. M., Port, A. M., McGuire, S., Ecker, A. J., Nasrini, J., Mollicone, D. J., Mott, C. M., McCann, T., Dinges, D. F., & Gur, R. C. (2015). Development and validation of the cognition test battery for spaceflight. *Aerospace Medicine and Human Performance, 86* (11), 942–952. https://doi.org/10.3357/AMHP.4343.2015.

Cohen, M. (2012). *Mockups 101: Code and standard research for space habitat analogues*. AIAA SPACE 2012 Conference & Exposition, AIAA 2012-5153, Pasadena, CA, 11–13 September 2012.

Cohen, M. M., Barker, D., Bianco, S., Bishop, S., Leon, P., Gentile, R., & Haeuplik-Meusburger, S. (2020). *Lunar daytime: Behavioral experiments in a space analog living and working environment*. International Conference on Environmental Systems, ICES-2020-011.

Cool Antarctica. (2001). *Fram – Ships of the Polar Explorers*. Retrieved July 1, 2020, from https://www.coolantarctica.com/Antarctica%20fact%20file/History/antarctic_ships/fram.php

Dempster, W. F. (1999). Biosphere 2 engineering design. *Ecological Engineering, 13*, 31–42.

Dudley-Rowley, M., Nolan, P., Bishop, S., Farry, K., & Gangale, T. (2001). Ten missions, two studies: Crew composition, time, and subjective experience in Mars-analog expeditions. In *Proceedings of the Third International Convention of the Mars Society*.

Eckart, P. (1996). *Spaceflight life support and biospherics*. Torrance, CA: Microcosm Press, Springer Science+Business Media Dordrecht.

ESA [Mars500]. (2012). *Mars 500 the isolation facility*. Retrieved May 20, 2020, from http://www.esa.int/Our_Activities/Human_Spaceflight/Mars500/The_isolation_facility

Haeuplik-Meusburger, S., Binsted, K., & Bassingthwaighte, T. (2017). *Habitability studies and full scale simulation research: Preliminary themes following HISEAS mission IV*. In 47th International Conference on Environmental Systems, Charleston, South Carolina, 16–20 July 2017.

Huntford, R. (1997). *Nansen*. London: Hachette Digital.

Kobak, A. (1999). *No man is a hero to his sled dogs, a biography of Fridtjof Nansen: Arctic explorer, Nobel Prize winner and mythical figure*. Retrieved June 11, 2020, from http://movies2.nytimes.com/books/99/05/16/reviews/990516.16kobackt.html

Nansen, F. (2011). *Farthest North – Being the record of a Voyage of Exploration of the Ship Fram, 1893–96, And of a Fifteeen Months' Sleigh Journey* (Vol. 1). Cambridge University Press.

NASA. (2014). *Human research program human exploration research analog (HERA) experiment information package*. Retrieved May 14, 2020, from https://www.nasa.gov/sites/default/files/files/HRP-HERA-Experiment-Information-Package.pdf (NASA [HERA] 2014).

NASA. (2015). *Human research program human exploration research analog (HERA) experiment information package*. Retrieved May 14, 2020, from https://www.nasa.gov/sites/default/files/atoms/files/hrp-nra-hera-study-information-package-06-2015_0.pdf (NASA [HERA] 2015).

NASA. (2016). *Human research program human exploration research analog (HERA) facility and capabilities information, July 2016*. https://www.nasa.gov/sites/default/files/atoms/files/july_2016_nra_hera_study_information_package_final.pdf. Accessed [date]. (NASA [HERA], 2016).

NASA. (2017). *Technology readiness level*. Retrieved December 10, 2019, from https://www.nasa.gov/directorates/heo/scan/engineering/technology/txt_accordion1.html (NASA [TRL] 2017).

NASA, ROI – Flight Analogs Human Research Program. (2019). *Human Research Program Human Exploration Research Analog (HERA) Facility and Capabilities Information*. Retrieved May 18, 2020, from https://www.nasa.gov/sites/default/files/atoms/files/2019_hera_facility_capabilities_information.pdf

National Science Foundation (NSF). (2015). *McMurdo Station Master Plan 2.1*, December 16, 2015. Accessed May 8, 2020, from https://future.usap.gov/master-plan/mcmurdo-master-plan-page/

Nelson, M. (2018a). *Lessons from biosphere 2, Both the scientific community and the general public still have a lot to learn from the largest mesocosm research project ever conducted.* Retrieved May 8, 2020, from https://www.the-scientist.com/reading-frames/lessons-from-biosphere-2-64464

Nelson, M. (2018b). *Pushing our limits: Insights from biosphere 2*. Tucson, US: The University of Arizona Press.

Nelson M., Pechurkin N. S., Allen J. P., Somova L. A., & Gitelson J. I. (2010). Closed ecological systems, space life support and biospherics. In L. Wang, V. Ivanov, & J. H. Tay (Eds.), *Environmental biotechnology. Handbook of Environmental Engineering* (Vol 10, pp. 517–565). Totowa, NJ: Humana Press.

Orlov, O. I., Belakovstky, M. S., & Ponomarev, S. A. (2015). *The moon of their own*. Retrieved May 15, 2020, from https://room.eu.com/article/A_Moon_of_Their_Own

Poynter, J. (2006). *The human experiment: Two years and twenty minutes inside biosphere 2*. New York: Basic Books.

Ryne, L. (2020). *Fridtjof Nansen: Man of many facets*. Retrieved June 11, 2020, from http://www.mnc.net/norway/Nansen.htm

Salisbury, F. B., Gitelson, J. I., & Lisovsky, G. M. (1997). Bios-3: Siberian experiments in bioregenerative life support. Attempts to purify air and grow food for space exploration in a sealed environment began in 1972. *Bioscience, 47*(9), 575–585.

Science Illustrated. (2012). *Dreaming of Mars*, part 1. Retrieved July 1, 2020, from https://scienceillustrated.com.au/in-the-mag/dreaming-of-mars-part-1/

Science.jrank.org. (2020). *Biosphere Project: Designing Biosphere 2*, (no date). Retrieved August 13, 2020, from https://science.jrank.org/pages/914/Biosphere-Project-Designing-Biosphere-2.html

Severinghaus, J., Broecker, W., Dempster, W., MacCallum, T., & Wahlen, M. (1994). Oxygen loss in biosphere 2. *Eos, Transactions American Geophysical Union, 75*(3), 35–37.

Sides, H. (2009). *1000 days in the ice. Norway's Fridthof Nansen was a pioneer of polar exploration*. Retrieved May 18, 2020, from https://www.nationalgeographic.com/magazine/2009/01/nansen/

Stuster, J. (1996). *Bold endeavors: Lessons from polar and space exploration*. Naval Institute Press.

The University of Arizona (UoA). (2020). *About Biosphere 2, Fast Facts, Biosphere 2 by the Numbers*. Retrieved May 18, 2020, from https://biosphere2.org/visit/about-biosphere2/fast-facts

Chapter 5
Review: Studies and Architecture of Habitability Missions in In-Situ Environments

This chapter continues the detailing of important milestones of the journey that lead to increasing the fidelity of habitability studies to eventually allow us to design the best-fit extraterrestrial habitats. Whereas Chap. 4 covered Mockups and Simulated Environments, this chapter gives an overview of exemplar **in-situ** (located in actual extreme environments) habitats and their associated research that have contributed to studies of extraterrestrial habitability and human factors. For each exemplar habitat, the physical and social settings, as well as mission relevant details, are summarized. Images and schematic plans illustrate the facility layout (see also Fig. 1.2). For readability and comparability reasons plans have been color coded according to primary human activities. Both short and long duration contributions to knowledge are briefly discussed to emphasize how habitability has slowly emerged as a central focus in the research on human adaptation and wellbeing in extraterrestrial living spaces. The chapter is similarly structured in alignment with the current NASA Technological Readiness Levels. Short guest contributions by astronauts and simulation crewmembers complement the information.

5.1 Habitability Studies in Terrestrial In-Situ Environments

The highest level of fidelity for habitability studies could only be achieved in fully realized modifiable facilities located in extraterrestrial locations. The compromise for terrestrial analogues is the utilization of an entire habitat and its associated facilities situated in real challenging environments with real risk and isolation aka 'in-situ'. The higher the degree to which terrestrial in-situ environments mirror critical aspects of extraterrestrial environments, (e.g., extreme isolation, risk, 100% closed loop environmental life support systems), the higher the fidelity for research purposes. NASA's TRL classification system identifies these as 'relevant

S. Häuplik-Meusburger, S. Bishop, *Space Habitats and Habitability*, Space and Society, https://doi.org/10.1007/978-3-030-69740-2_5

environments'. Terrestrial in-situ facilities are considered hybrids, offering differing degrees of risk and isolation as well as variable crew autonomy but with lesser control over extraneous factors than would be inherent in a laboratory mockup or simulation facility. Most do not represent completely closed loop life support systems; an exemplar exception would be underwater facilities.

With reference to the Technology Levels used at NASA, this group encompasses the TRL Levels 5–6:

- TRL 5—Component and/or breadboard validation in laboratory equipment
- TRL 6—System/subsystem model or prototype demonstration in a relevant environment (ground or space)

5.1.1 Underwater Research Laboratories

Underwater facilities offer means for testing technologies, systems, tools, and operations of the habitat and EVA-related activities. They are also the only analogues where crew operations, equipment, tools and vehicles in simulated microgravity conditions can be tested over longer periods (in addition to the short tests in neutral buoyancy pools).

5.1.1.1 Conshelf I–II

Although not technically considered a research facility, it is worth noting that the first underwater habitat was Conshelf I (Continental Shelf Station) in 1962 off the coast of Marseilles in France situated at a depth of 10 m (32.81 ft). Conshelf I was the first of a series of three underwater habitats planned as an 'underwater village' by Jacques Cousteau, a famous French pioneer in marine conservation and underwater filming. He wanted to demonstrate that it was possible to live underwater for longer periods of time. His colleagues, Albert Falco and Claude Wesley lived in the cylindrical module with 5 m (16.4 ft) in length and 2.5 m (8.2 ft) in diameter for 1 week. It was followed by Conshelf II in 1963, in the Red Sea. Conshelf II consisted of three facilities, the starfish-shaped habitat, a submarine garage and a laboratory at 25 m (82 ft) in depth. Five people lived for 1 month in the Starfish habitat, two more in the Deep Space laboratory. Conshelf II is documented in Cousteau's documentary film "*World without Sun*" (1964).

5.1.1.2 Tektite I–II

The Tektite habitat was an underwater laboratory located in Great Lameshur Bay, Saint John, U.S. Virgin Islands. The research program dealt with aspects of coral reef ecology and "*to show that saturation diving from an underwater laboratory could be done efficiently, safely, and at a relatively small cost*" (Jarvis 2016). The nationally

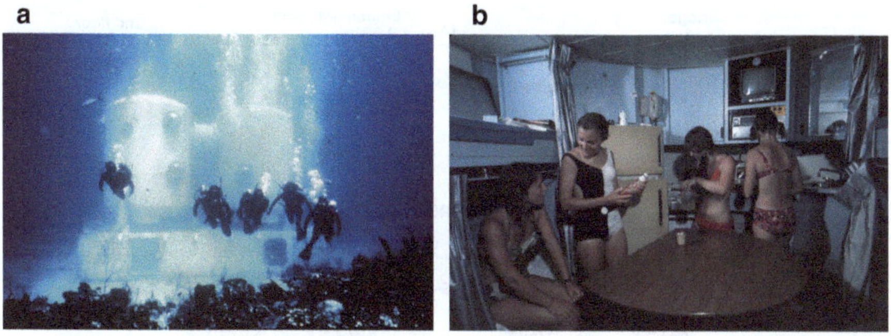

Fig. 5.1 (**a**) The Tektite habitat in 1969 with the aquanauts (Credit: OAR/National Undersea Research Program-NURP). (**b**) The Mission 6 all-female crew during Tektite II in 1970 with Sylvia Earle in b & w suit (Credit: OAR/National Undersea Research Program—NURP)

sponsored programs Tektite I and Tektite II included nine studies and took place in 1969 and 1970 (see Fig. 5.1a, b).

5.1.1.2.1 Facility Architecture and Design

The underwater habitat was designed and constructed by the General Electric Company and was furnished to the Navy under a bailment agreement[1] (Pauli and Cole 1970, p. 10). It laid at a depth of 15 m (49 ft) with the habitat hatch about 2 m (6 ft) above the bottom. Tektite I consisted of a rectangular base of 15 m (49 ft) length and two metal cylinders. The cylinders were 3.8 m (12.5 ft) in diameter, 5.5 m (18 ft) high and connected with a flexible tunnel. Air, water and electric power was provided through multiple cords from the shore.

The cylinders had two stories (see Fig. 5.2) and were divided into two compartments each: bridge, crew quarters, equipment room, and wet room. The *wet room* in the bottom of one of the cylinders served as entry and exit for the aquanauts through a ladder down to the base. It also served as wet laboratory and storage for scientific and diving equipment. Above the wet room was the equipment or engine room. It contained the life-support systems, a freezer for food storage, and a bathroom with sink and toilet. The cupola was above. The bridge or control room was accessed via the flexible tunnel. It served as control center and dry laboratory. The habitat engineer slept here on a folding cot. The crew quarters were in the bottom of this cylinder and contained four bunkbeds, storage space for personal gear, and radio and television. They had a dining table in the center, as well as kitchen appliances and storage space. The interior included carpeted floors.

[1]A **bailment agreement** is where one person agrees to take physical possession of another person's property for some purpose, but does not take ownership of it, with the understanding it will be returned at a later date. A ***bailment*** involves only a transfer of possession or custody, not of ownership. A rental or lease of personal property might be a ***bailment***, depending upon the ***agreement*** of the parties. https://legal-dictionary.thefreedictionary.com/bailment.

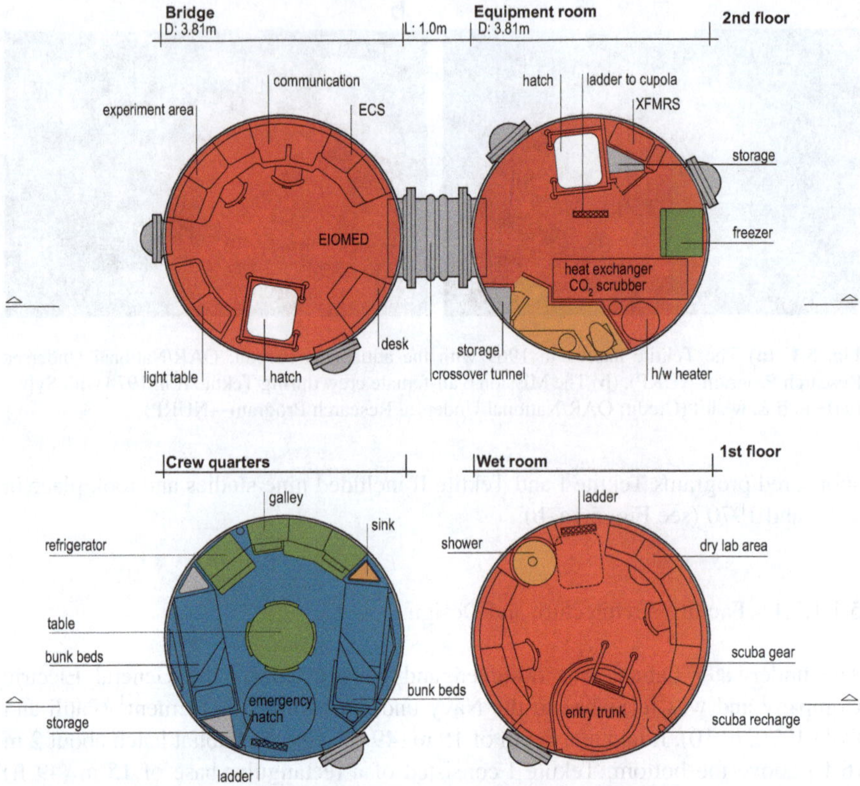

Fig. 5.2 Schematic plans and section of the underwater habitat Tektite (Credit: The Authors)

5.1.1.2.2 Habitability Studies, Experiences and Lessons Learned

Tektite I was coordinated by the Office of Naval Research. Much of the research for Tektite I centered on human adaptation in this new environment (Reichard 2010). Four scientists (Ed Cliffton, Conrad Mahnken, Richard Waller and John Vanderwalker) worked and lived in the Tektite habitat for 60 days. Topics investigated were marine science, life sciences behavioral, life sciences biomedical, and ocean engineering. Marine science included ecological, biological and geological studies in-situ, as well as an evaluation of diving techniques. The life science behavioral studies investigated "*crew size and selection criteria, quarters size and habitability, and use of time in mission performance*" (Pauli and Cole 1970, p. 15). The experiments included measures on general activity, task performance, efficiency, social relations, operational and interpersonal communications, personal habitats, emotional adjustment, psychological traits, psychomotor performance, and sleep (Pauli and Cole 1970).

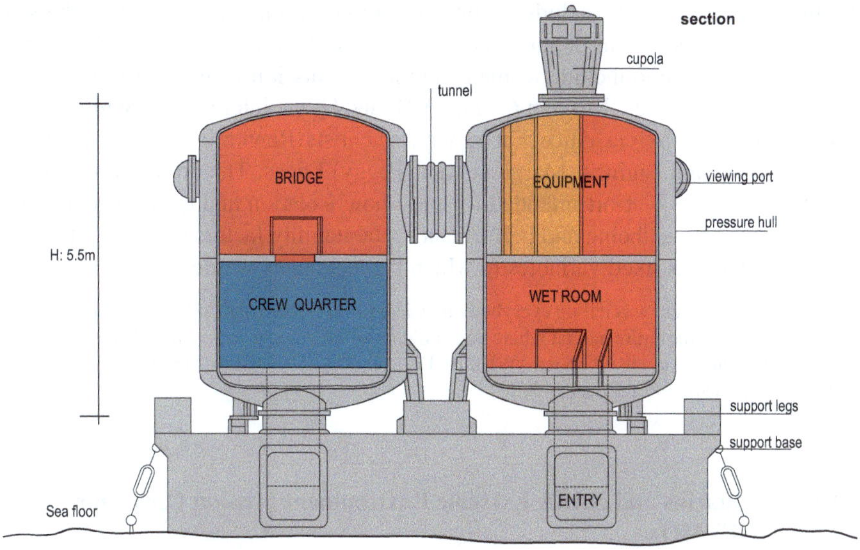

Fig. 5.2 (continued)

In addition, individual and group responses to habitability factors (living space, arrangement of work and rest areas, lighting and recreational facilities) was of interest. Behavioral studies focused on crew relationships, crew performance, psychological state of mind, stress responses and coping. Many of the research results confirmed the absence of negative or deleterious effects from living at depth. A major conclusion was that 2 months was simply not long enough to adequately evaluate long duration risks. This realization was a major factor in the design of subsequent missions.

The life science biomedical studies investigated medical safety and physiological effects of long-term saturated diving (oxygen toxicity, hematology, microbiology, respiratory as well as decompression effects). The engineering studies focused on ocean engineering, entailing the design and fabrication of the habitat and evaluating its performance, in "*that the habitat as a whole provided a comfortable and livable home for the Tektite I long-duration, shallow water, saturated dive. As a laboratory it was not optimum, but it was adequate.*" (Pauli and Cole 1970, p. 28).

Tektite II was coordinated by the Department of the Interior and comprised 10 missions, which lasted between 10 and 20 days with a crew of four scientists and one engineer. In addition to the research conducted inside the habitat, the aquanauts spent up to 12 hours in the water each day. NASA undertook a Habitability Research Program to "*generate research data on human behavioral dynamics and habitability assessment for small crews in confined living spaces*" (NASA [Tektite II] 1972, p. I-1). During the course of the project, over 60 scientists and engineers lived and worked beneath the sea in addition to the numerous teams of scientists, professors, students and other participants observing and studying the aquanauts from land.

Tektite set many records including: the first long-term scientific mission in the sea, the first that undertook in-depth ecological studies, the longest saturation dive, the first NASA mission to include women and the first mission with an all-female crew who became famous as 'Mission 6' (Fig. 5.1b) led by the internationally recognized marine biologist, Sylvia Alice Earle, with scientists Renate True, Ann Hartline, Alina Szmant, and engineer Margaret Ann (Peggy) Lucas. The project was considered the first NASA effort intended to study how isolation and confinement could impact astronaut wellbeing and performance. Habitability factors before Tektite had been considered as fixed variables to which the aquanauts would have to adapt.

> The hardest moment was telling everybody they had to slow down and do things right. There was one spot in the entire habitat where you could go without being seen or heard, and two of us went in there and talked about, "What do I do next?" [...] "That was the worst". Peggy Lucas Bond (Spaceflight Insider 2013).

5.1.1.3 Aquarius and NASA Extreme Environment Mission Operations (NEEMO)

The Aquarius research station is the only underwater analogue research facility still currently active. It is located 18 m (62 ft) underwater, 3.5 miles off Key Largo in the Florida Keys National Marine Sanctuary. Aquarius was built in 1986, first deployed in 1988 in the U.S. Virgin Islands and finally installed in Florida in 1992. It is owned and operated by the National Oceanic and Atmospheric Administration (NOAA). The research station has been used by multiple organizations and universities from the US as well as from other countries. Since 2001, Aquarius is used for the NASA Extreme Environment Mission Operations program (see Fig. 5.3a, b). The scientists

a b

Fig. 5.3 (**a**) NEEMO 11 crewmember works near the undersea habitat "Aquarius" during a session of extravehicular activity (EVA) for the NASA Extreme Environment Mission Operations (NEEMO) project (Credit: NASA). (**b**) The crew taking part in NEEMO 22, the 22nd NASA Extreme Environment Mission Operations mission, consisted of astronauts, technicians and scientists. NASA astronaut Kjell Lindgren, commander is joined by ESA astronaut Pedro Duque, planetary scientist Trevor Gradd and research scientist Dom D'Agostino, along with two support technicians (Credit: NASA Analogs)

"*work to understand the changing ocean and the condition of coral reefs, which are threatened locally, regionally and globally by increasing amounts of pollution, over-harvesting of fisheries, disease and climate change*" (NASA [Aquarius] 2006).

5.1.1.3.1 Facility Architecture and Design

The Aquarius facility has three elements: a life-support buoy floating on the surface, the Aquarius habitat, and the baseplate that holds it to the ocean floor (NASA [Aquarius] 2006). The 14 m long (45 ft) and 3 m (9.8 ft) in diameter sized facility hosts about 41 m^2 (440 ft^2) living and working space. Two waystations with pockets of air for divers to take a break, complement the facility. Access to the water is provided through a 'wet porch', similar to an inverted glass that is half filled with water and half filled with air. Next to it is the hatch (see Fig. 5.4). Aquarius is equipped with sophisticated lab equipment and computers, enabling scientists to perform research and process samples without leaving their underwater facilities.

The Aquarius facility requires total dependency on life support with "*significant restrictions to escape or access to immediate help*" (Bishop 2011, p. 70). This is because the Aquarius underwater habitat is an "*ambient pressure habitat, which means that the interior atmospheric pressure (about 2.6 atmospheres) is equal to the surrounding water pressure*" (NASA [Aquarius] 2006). Stays for longer than 80 min require a 17-h decompression activity in the habitat before a return is possible to the ocean surface.

5.1.1.3.2 Habitability Studies, Experiences and Lessons Learned

Aquarius provides a capsule habitat uniquely situated within an environment that replicates many of the closed-loop constraints of the vacuum of space, i.e., a hostile, alien environment that requires total dependency on life support; poses significant restrictions to escape or access to immediate help; has limited supplies of critical resources and is defined by limited, confined habitable space and physical isolation. The complexity of NEEMO missions further parallel space missions in their mission architecture, with similar requirements for extensive planning, training, control, and monitoring via an external mission control entity. However, it has only been the most recent NEEMO missions in which stress, fatigue, and cognitive fitness, as well as individual and intrapersonal mood and interaction, have been the focus of study. Interestingly, a habitability survey including questions regarding the acceptability of the habitat was conducted during missions 1–5, with a live feed from cameras added for NEEMO 5 (cf. Häuplik-Meusburger et al. 2017).

Examples of the resulting video analysis found multiple instances where habitability improvements were needed (NASA NEEMO-HABITABILTY 2004):

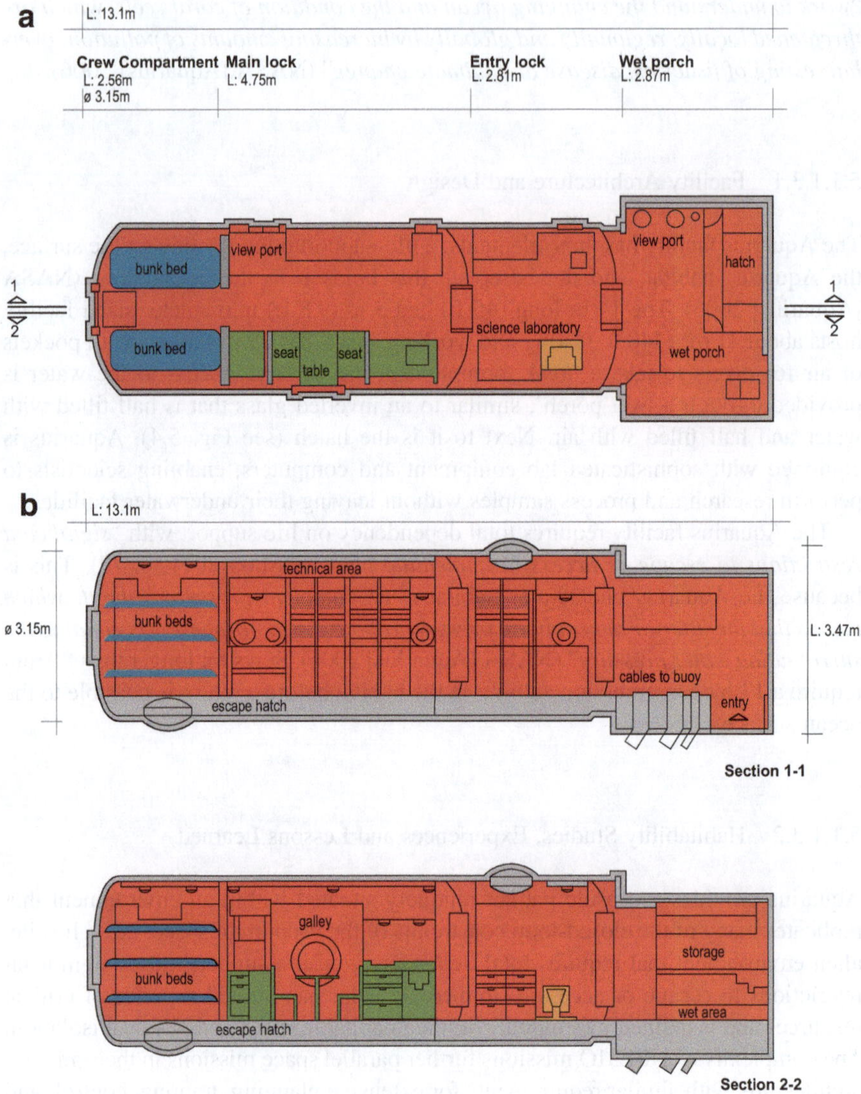

Fig. 5.4 (**a**) Schematic floorplan of Aquarius (Credit: The Authors). (**b**) Schematic sections of Aquarius (Credit: The Authors)

- Workstations were poorly placed and interfered with the flow of traffic in and out of the Main Lock. Particularly, the two workstations located near the entrances were problematic both for people trying to pass through, as well as for people working at them who were interrupted in their work.

- The table was an inappropriate size. Used for eating, working, meetings, and group teleconferences with mission control or temporary visitors, the table sometimes needed to accommodate up to six people at once. Even for four people, as was more often the case, the table was far too small, and it was obvious from the video that the crewmembers were uncomfortable.
- The food storage was poorly sited. Multiple instances were recorded where crew members were seen reaching across the small galley, blocking traffic, or standing precariously on the stepladder to reach the storage areas.
- The refrigerator and microwave/storage areas were located too far from each other, as lots of back-and-forth traffic occurred between them. Given the narrow passage, it was readily apparent that traffic within the Main Lock could be much ameliorated simply by moving these two areas closer together.

5.1.2 Antarctic Research Stations

Antarctica has been acknowledged as the analogue environment with the highest fidelity for space primarily for the extremity of the environmental risk, degree of isolation during the winter months and the extent of dependence upon resources that must be provided. The lived experience of Antarctica is believed to be the closest analogue to the stress that future long duration space crews will also deal with. In addition to the fear of illness or injury in a place far from extensive medical facilities and prolonged periods of not being rescued or resupplied, extreme environmental conditions (high winds, extraordinary low temperatures, low levels of oxygen, extreme dryness, extended periods of day/night) demanding persistent technological support which could fail, there's also the psychosocial factors of the monotony of eating, sleeping, working and spending leisure time in one place with the same people for months on end.

> If you're stuck there with somebody you really can't stand, too bad. You're stuck with them. And if you're missing somebody who's far away, too bad. You're stuck without them [Peter Suedfeld (Bennett 2016)].

While there are approximately 75 bases in Antarctica (of which approximately 40 are occupied year around) (Cool Antarctica, https://www.coolantarctica.com/Community/antarctic_bases.php), there are only three permanent, all-year purely research stations on the Antarctic Plateau: Vostok Station (Russia), the Amundsen-Scott South Pole Station (U.S.) at the geographic South Pole and Concordia (France/Italy) on Dome C. All the rest are considered operational stations with national and exploratory as well as research mandates. Those three will be introduced in the following sections.

5.1.2.1 Amundsen-Scott South Pole Station

The Amundsen-Scott South Pole Station is a U.S. scientific research station located at the South Pole, the southernmost place on the Earth and the largest desert on Earth (Geoscience News and Information 2020) found on the Antarctic high plateau 2835 m (9,301 ft) above sea level (see Fig. 5.5a, b). Built in 1956–57, it was the first station specifically intended to support research of the polar regions of Earth during the International Geophysical Year and has been continuously occupied, rebuilt, expanded and upgraded numerous times. Over the years, researchers have studied geophysics, glaciology, meteorology, upper atmosphere physics, astronomy, astrophysics and biomedical factors. The station is the only inhabited land location which has 6 months of constant daylight and then 6 months of constant darkness. The summer population is ~150 with less than 50 remaining over the polar winter who are then completely isolated between mid-February to late October (Box 5.1). In addition to the scientists, there are cooks, plumbers, doctors, internet technicians, electricians and other support staff needed to keep the station running.

5.1.2.1.1 Facility Architecture and Design

The preliminary station 'Old Pole' was made of prefabricated elements, which were airdropped and put together by the US Navy Seebees in 1957. Over time the station was buried with snow and was abandoned in 1975. The second facility was operational from 1975–2003, within a geodesic dome with a diameter of 50 m (160 ft), which also was slowly covered by snow. It was dismantled in 2009, shortly before the latest facility became operational.

Begun in 2003 and completed in 2010, the latest version of the station sports an innovative modular two-story adjustable elevated design that utilizes the Antarctic wind flow to scour the accumulating snow away from the base to prevent its

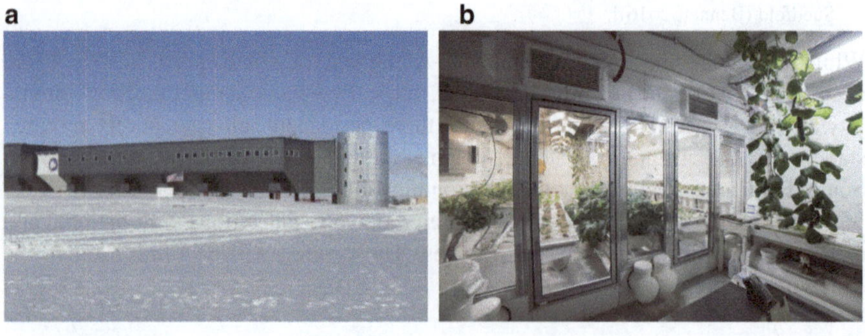

Fig. 5.5 (**a**) Exterior view of Amundson-Scott South Pole Station (Credit: NSF/USAP, photographer Amy Lowitz). (**b**) Hydroponic greenhouse at South Pole Station, 2016 (Credit: NSF/USAP, photographer Amy Lowitz)

submersion.[2] Similar to other recent arctic stations its architecture features a high-performance insulation skin as well as heat energy conservation appliances.

Amundsen-Scott South Pole Station encloses 7400 m^2 (80,000 ft^2) including a small hydroponic greenhouse producing the only source of fresh fruit and vegetables during the winter as well as a gym, movie room, music room, rock-climbing wall, library, cafeteria, store, single-person bedrooms (with internet connections) and a sauna. Day-to-day station operations (science, medical, administration, food service, berthing, communications) are located primarily on level 2 while crew accommodation, gymnasium and support spaces (mechanical, electrical, emergency power generation) are situated on level 1 (see Fig. 5.6). The entire elevated complex (exclusive of the remote science facilities) is connected to a below surface main power plant, cargo facility, garage/shop, and fuel storage by means of a vertical circulation tower, which includes a staircase, cargo lift, and all plumbing and utility risers (Ferraro Choi and Associates Ltd. 2017). Access to the internet (and contact with the outside world) is limited to the 2–4 h when communication satellites pass overhead.

5.1.2.1.2 Habitability Studies, Experiences and Lessons Learned

Most of the habitability research conducted at South Pole Station has focused on the application and testing of bioregenerative life support technology and recycling systems intended to meet the mandate to preserve the Antarctic environment while addressing NASA's goals of developing regenerative food, water and waste-processing systems for long duration space missions. In 2016, the National Science Foundation (NSF) (who oversees all U.S. research stations and vessels in Antarctica) and NASA signed a collaborative agreement to support human health research in polar environments finally incorporating advanced habitability of ICE environments as one of the primary foci (NSF 2016).

There are multiple factors that make the South Pole station unique. For a start, as the only inhabited place on the surface of the Earth from which the Sun is continuously visible for 6 months and is then continuously below the horizon for the next 6 months, there is essentially one long day and one long night with 2 days between of twilight (i.e., the equinoxes) . . . for the entire year! The geographic position of the South Pole means you can literally travel through all 12 time zones by walking in a small circle. They even have developed a Christmas tradition where inhabitants race around a two-mile course in various vehicles to complete a 'race around the world' each year.

[2]The station has 36 hydraulic jack columns intended to allow corrections for differential settling and accommodation for eventual snow build-up. It is the fourth elevated base to have been built; see Table 1 in Ferraro Choi and Associates Ltd. (2017) https://ferrarochoi.com/publications/elevated-station-design/ for details on other elevated bases. Other elevated polar stations are the Concordia station, Brazil's Comandante Ferraz Research Station (2020), Britain's Halley VI (2005), India's Bharati Station (2012) and Spain's Gabriel de Castilla (2018).

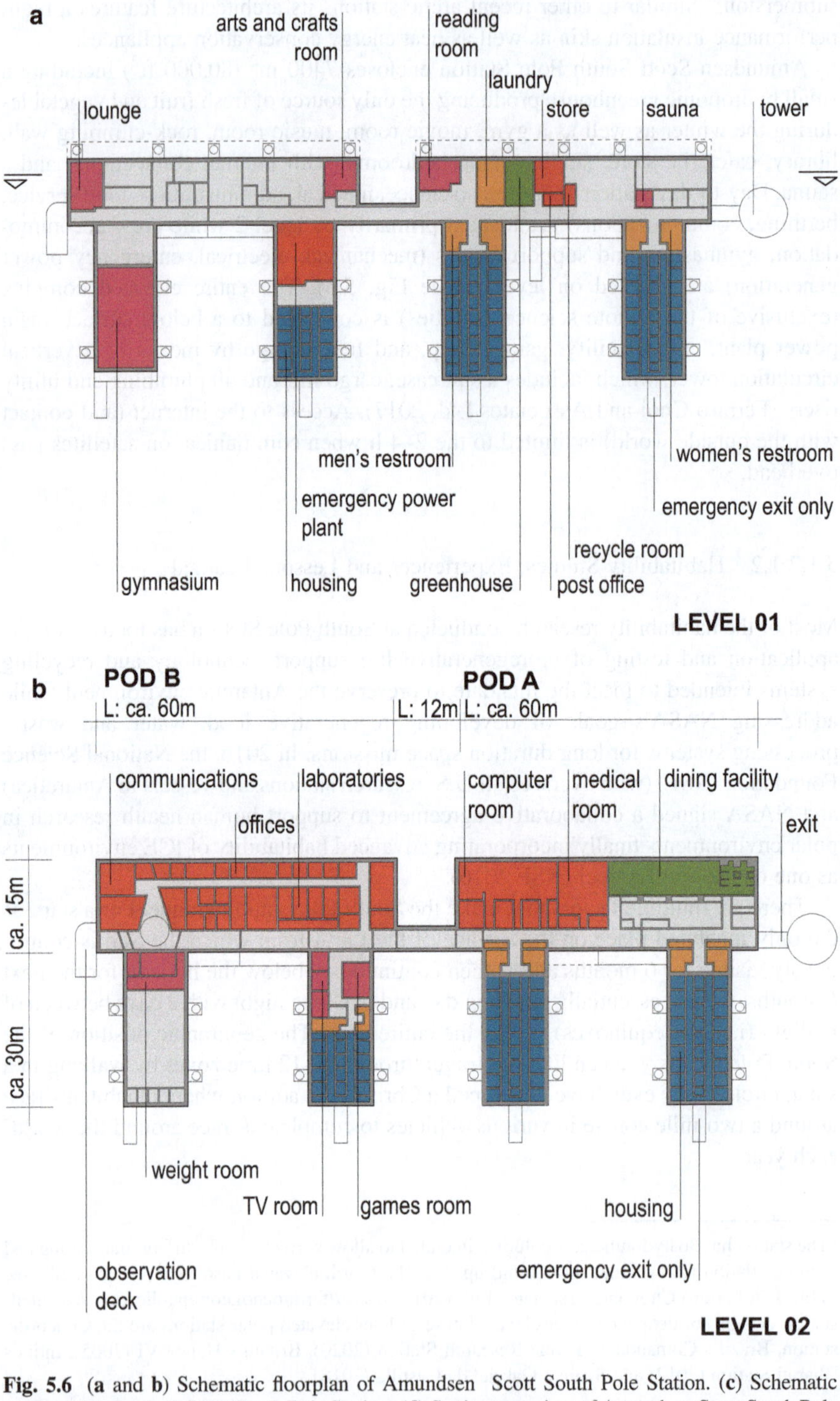

Fig. 5.6 (**a and b**) Schematic floorplan of Amundsen Scott South Pole Station. (**c**) Schematic section of Amundson Scott South Pole Station. (**d**) Station overview of Amundson Scott South Pole Station (Credit: The Authors)

c

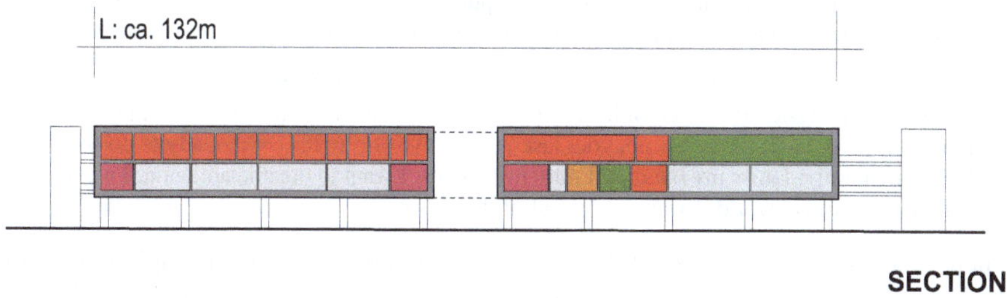

d

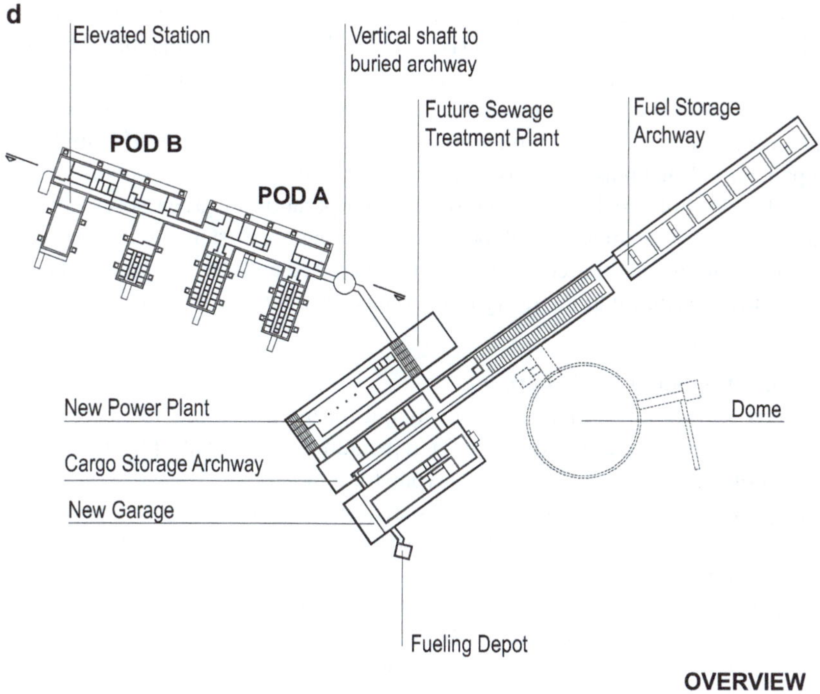

Fig. 5.6 (continued)

Wintering-over presents notorious dangers and stresses, as the station population is almost totally isolated. The isolation is so extreme that there have only been three medical evacuations in the history of the base and even those were dependent on breaks in the weather and favorable conditions becoming available. Most famous was the 1999 case in which a doctor for South Pole Station discovered she had breast cancer and treated herself for almost 6 months, performing her own biopsy and administering her own chemotherapy until a rescue plane was cleared for takeoff (Box 5.1). Then, in 2016, Astronaut Buzz Aldrin, who walked on the moon, visited South Pole station as a tourist, and fell ill and had to be medically evacuated. At age 85, he was the oldest person to have visited the South Pole.

> "There's a pretty consistent trend that when you land there you're bright-eyed and bushy-tailed and you're happy and you're ready to put in all this effort and work," said Tyler St. Germaine, who has visited the station for two-month stays four times in as many years.
>
> "You're full of energy and you're full of optimism," said St. Germaine, who earned his Ph.D. in astrophysics at the end of his rotation in Antarctica. "But almost every time, by the time your six or eight weeks are up, your last week or two you're like, 'I'm ready to go home. I miss my family. I miss my internet. I miss my friends. I want to breathe air again. I want to experience the real world. I want to get out of this place.'"[3]

Then, there is the greenhouse at the station. Greenhouses have been implemented in Antarctic field sites since 1902 with more than 46 distinct plant production facilities, at one time or another, in operation in Antarctica. A review of Antarctic greenhouses and plant production facilities in 2015 (Bamsey et al. 2015) noted that the "need for this biological association can be solely psychological while at other times it is based on the more practical need for fresh food." The variety of vegetables and herbs in the greenhouse range from fresh eggplant to jalapenos and are all produced completely hydroponically. What makes the South Pole greenhouse special is the opportunity to explore the psychological importance of the incorporation of green spaces by the biomedical research team as part of their research program. Systematic investigation into the role and impact of green spaces in impoverished and ICE environments will provide critical insight into the design of habitats for future space habitats.

Box 5.1 I'm going to die here

Dr. Jerri Nielsen knew there was no way out for the next 8 months, because planes could not land due to the low temperatures. As the only physician at South Pole Station, she had not told anyone for 3 months since she discovered a lump in her breast. Instead, she had started training the other crew to take over for her. As the lump grew bigger, she told the head of the station, and chemotherapy was arranged for her.

Later in her book she would write: "*I can't decide if I am frightened or relieved. I am sick and quite possibly dying [...] And how I was not afraid. Here, in this lonely outpost surrounded by the staggering emptiness of the polar plateau, in a world stripped of useless noise and comforts, I found the most perfect home I have ever known. I do not want to leave.*" (Nielsen and Vollers 2001)

Read more in the book Ice Bound: A Doctor's Incredible Battle for Survival at the South Pole (2001)

5.1.2.2 Vostok Station

The Russian research station, Vostok, was founded in 1957 in the inland Princess Elizabeth Land region of Antarctica at the Southern Pole of Cold with the lowest reliably measured natural temperatures on Earth (−89.9 °C (−128.6 °F; 184.0 K)

[3]https://news.harvard.edu/gazette/story/2019/09/harvard-scientists-at-the-south-pole/.

Fig. 5.7 Panoramic photo of Vostok Station. The striped building on the left is the power station while the striped building on the right is where researchers sleep and take meals. Building in the background with the red- and white-striped ball on top is the meteorology building (Credit: NOAA, photograph Todd Sowers LDEO, Columbia University, Palisades, New York)

and the South Geomagnetic Pole. Research has focused on ice core drilling and magnetometry with additional studies in actinometry,[4] geophysics, medicine and climatology. Located at 3488 m (11,444 ft) above sea level, it is one of the most isolated established research stations in Antarctica. Hosting ~25 scientists and engineers in the summer, the winter-over crew typically numbers 13 members.

The station became well known for the discovery of the largest freshwater lake in the world lying some 4000 m (13,000 ft) below the surface of the central Antarctic ice sheet and covering an area of 14,000 km^2 (5400 mi^2).

5.1.2.2.1 Facility Architecture and Design

Named after the ship of the First Russian Antarctic Expedition, Vostok is composed of five main aluminum frame buildings interconnected by tunnels and covered with snow. There is also a glacial-drilling complex of veneer panels with heat insulation, an emergency electric power station and a base of fuel lubricants (see Fig. 5.7). The aging station is in poor condition with few upgrades (see Fig. 5.8a, b).

[4]Chemical actionometry involves measuring radiant flux via the yield from a chemical reaction, to estimate the temperature of the Sun's surface.

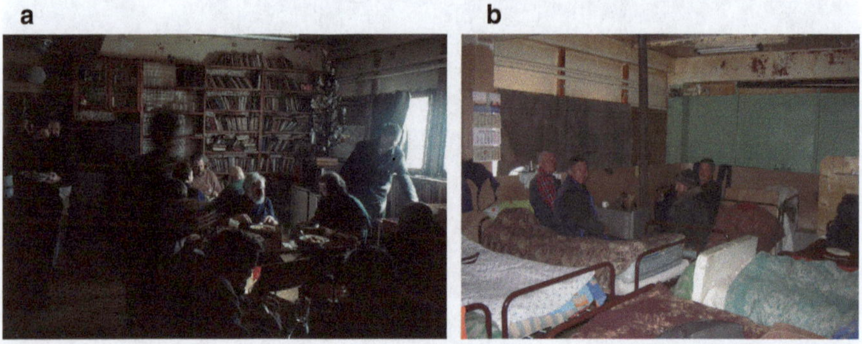

Fig. 5.8 (**a**) Vostok Galley (Credit: Alexey A. Ekaykin). (**b**) Vostok Sleeping Quarters (Credit: Alexey A. Ekaykin)

5.1.2.2.2 Habitability Studies, Experiences and Lessons Learned

Multiple factors make Vostok one of the most difficult places on Earth for human habitation: an almost complete lack of moisture in the air, an average windspeed of 5 m per second (11 mph: 18 km/h), a lack of oxygen because of the high elevation, a higher-than-normal ionization of the air and a polar night that lasts 120 days. These conditions require significant acclimatization that takes a week to 2 months. The process of acclimatization may involve headaches, eye twitches, ear pains, nose bleeds, feelings of suffocation due to the low oxygen content, precipitous rises in blood pressure, insomnia, vomiting, loss of appetite, joint and muscle pain, arthritis and weight loss.

The difficult conditions constitute considerable stress for crews. The station Vostok became infamous for a 1959 altercation between two scientists over a game of chess which resulted in an attack and injury which may or may not have resulted in the death of one member (accounts differ), possibly making it the first Antarctic base to have to deal with a murder of one crewmember by another (Bennett 2016; Joyner and Chopra 1988, p. 67; Barrett and Martin 2014, p. 88).

5.1.2.3 Concordia Research Station

Completed in 2005, the Concordia research station is a French-Italian base, sited on the Dome C, one of the three major peaks of the Antarctica Plateau, about 1800 km from the South Pole. The research station consists of two cylindrical towers and a technical building interconnected by enclosed walkways forming the permanent station and a summer camp, with containers and tents (see Fig. 5.9a, b). In winter, the facility can house up to 16 persons in complete autonomy for 9 months. In the summer, the population will double to 32–40 residents. Concordia is the remotest

Fig. 5.9 (**a**) Concordia image outside (Credit: IPEV/PNRA/ESA–S. Thoolen). (**b**) Dinner time inside Concordia station (Credit: IPEV/PNRA/ESA–S. Thoolen)

base on Earth, as "*during the harsh winter no outside help can be flown in or reach the base over land*" (ESA 2013).

5.1.2.3.1 Facility Architecture and Design

The permanent station is composed of two towers with three floors each (Candidi and Lori 2003, p. 5) resting on six hydraulic legs that can be raised to adapt to changes in topography and prevent snow-drift accumulation. One building contains life support and logistic services (e.g., kitchen, dining, living rooms, workshops) and is considered 'noisy' while the other contains the living areas (bedrooms and laboratories) and considered 'quiet' (Fig. 5.10a, b). The total area is about 1500 m^2 (16,146 ft^2).[2]).

5.1.2.3.2 Habitability Studies, Experiences and Lessons Learned

Concordia was specifically designed as an analogue facility to study human adaptation to isolated, confined extreme environments in addition to the projects conducted by the crew on climatology, meteorology, astronomy and geophysics. The winter-over period at Concordia is longer with 9 months of darkness and isolation compared to the 6 months at McMurdo. "*For those nine winter months, the crew will experience extreme isolation. Concordia is a real operational environment, something we would never be able to simulate in a laboratory. This will enhance and complement our research and give us valuable insight we need to prepare for Mars*" (ESA 2005).

Human presence in Concordia is totally dependent on life support systems (including temperature and heating, constant artificial lights). Living conditions in Concordia involve sensory, intellectual/work, recreational and social monotony. Leisure activities in the station are sparse (Box 5.2). The extreme separation and

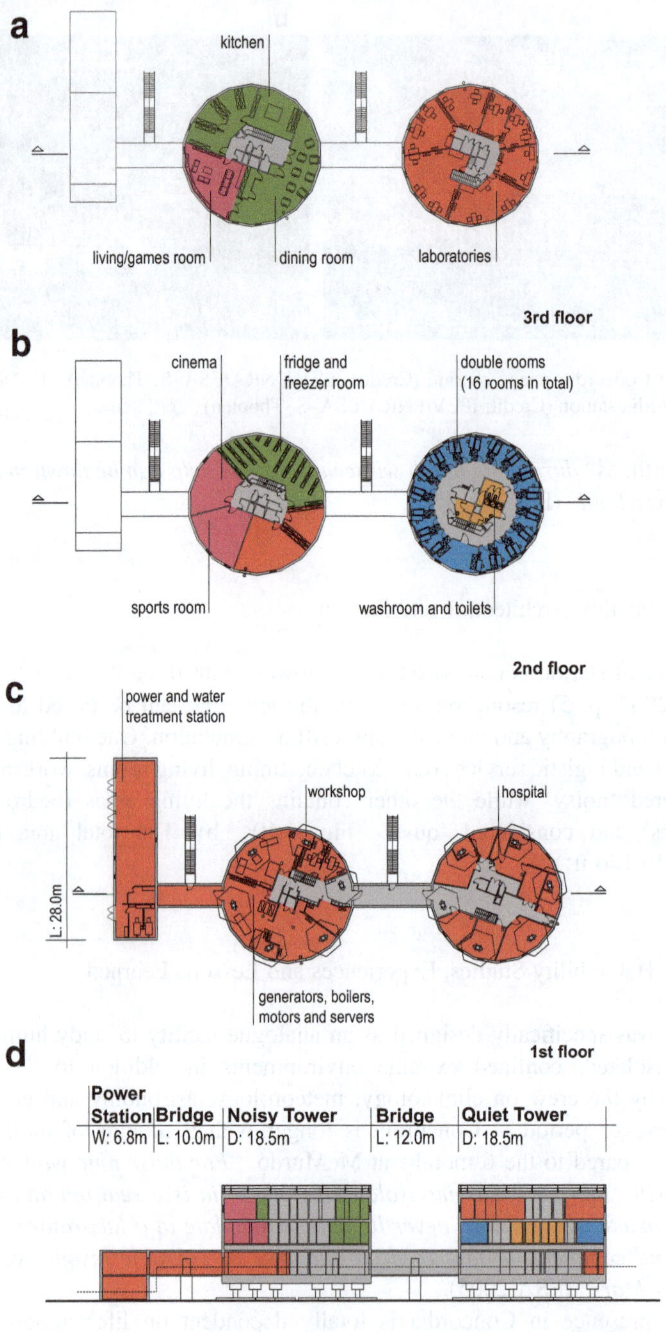

Fig. 5.10 **(a–c)** Schematic plans of Concordia Research Station (Credit: The Authors). **(d)** Schematic section of Concordia Research Station (Credit: The Authors)

isolation lead to limited telecommunications and poor access to information from the outside world, including communication with friends and family. However, each participant has a private room with their personal belongings. Food is generally well prepared by a cook, although it is from frozen and dried ingredients (Nicolas et al. 2016). Research from Concordia has indicated that preferential uses for living space may be driven by different social and occupational dimensions with scientists expressing a greater need for privacy while technicians prefer the social leisure area (Weiss et al. 2007). Such early findings reinforce the importance of characterizing the experience of various living spaces in order to understand the utility and value of such spaces.

Box 5.2 A large indoor space is missing

The station is extremely well planned for the amount of space we have, for 13 people it is exceptionally good. The only thing I would say that is missing here, that we all miss, is a [larger] space inside. We have a gym room, but no space where we could play football, basketball or volleyball. It would be great to have a building like that. I realized that the South Pole Station has such a facility. I think that would add a lot to our environment, especially in the deprivation we are undergoing.

Source: From a personal interview with Alexander Kumar while overwintering at Concordia research station in July 2012. A radio broadcast on that interview can be listened here: https://cba.fro.at/65695[5]

5.1.3 Other Terrestrial Facilities

There are other existing terrestrial facilities with the goal of simulating space missions. Most of them failed to meet our requirements to study habitability (see also Chap. 1). The following facilities exemplify relevant characteristics that are interesting and contribute to the knowledge on habitability in extreme environments.

5.1.3.1 Haughton-Mars Project (HMP)

One of the first of dedicated research hybrid facilities was the Haughton-Mars Project (HMP). It was initiated in 1996 when the National Research Council of the U.S. National Academy of Sciences and NASA Ames Research Center sponsored a postdoctoral proposal to study the Haughton Crater on Devon Island, the world's largest uninhabited island, in the Canadian Arctic as a potential analogue for Mars. Its harsh climate with freezing temperatures, isolation and remoteness makes it a

[5]Kumar 2012, interview, radio broadcast, Häuplik-Meusburger, S., Holzgethan, V. (Host). (2012, October 9). *Alexander Kumar gives an insight to daily life in Antarctica*. In Spaceuriosity. https://cba.fro.at/65695.

unique analogue for simulating challenges for future Mars missions (NASA [HMP] 2019).

The program has expanded from a four-member team in 1997 to a permanent habitat that hosts eight-week arctic summer field seasons with 50 to 90 participants, multiple teams, and research projects that run from instrument testing and development to biomedical and psychological evaluation. HMP routinely supports participation by NASA; the Canadian Space Agency (CSA); the Russian Institute for Space Research (IKI); various research institutions and universities in the United States, Canada, and the United Kingdom; and the U.S. Marine Corps. It has been the subject of various documentaries made by such groups as the National Geographic Society and Discovery Channel Canada (Mars Institute 2007).

5.1.3.1.1 Facility Architecture and Design

The HMP Research Station (see Figs. 5.11 and 5.12) includes a main base and a series of modular tents radiating next to the core, including the Office, Systems, Geology Lab, Astrobiology Lab, Garage/Workshop, and Medical Tents. The Arthur Clarke Mars Greenhouse (ACMG) is an additional stand-alone structure supporting studies in plant growth research. The HMPRS also includes the HMP Airstrip, a 200 m (800 ft) long dirt strip. However, the Haughton Mars Project (HMP) is a

Fig. 5.11 Haughton-Mars Project Research Station, Devon Island, 2006. As a whole, the base can currently accommodate about 40 to 50 people at a given time, with researchers sleeping in individual tents near the main structures (Credit: Matthew Silver)

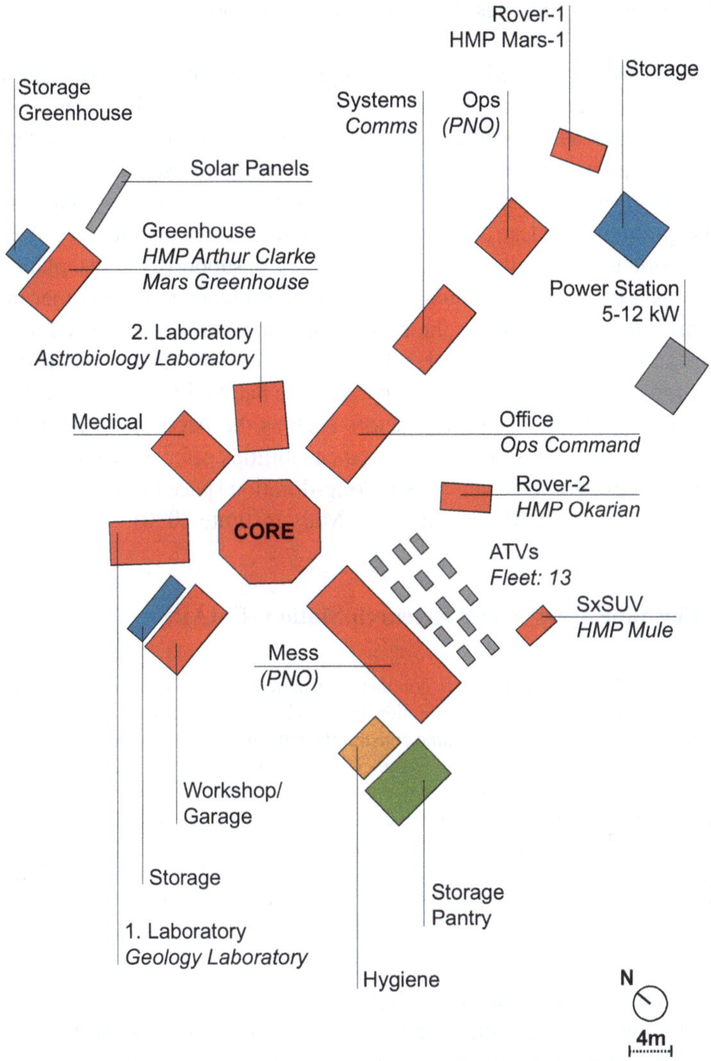

Fig. 5.12 Schematic plan of the Haughton-Mars Project (Credit: The Authors)

"*bring your own tent*" for field participants with each participant expected to bring his/her own personal camping tent and sleeping bag. An "Equipment List", updated each year, is provided to each confirmed participant to allow him/her to plan for all needed gear.

The HMP operates a substantial fleet of surface mobility systems, including 12 ATVs (quad bikes), 1 Side by Side, 1 Amphibious Rover, and 2 Humvees. The Humvees serve as analogue pressurized rovers, as well as mobile field labs, mobile field shelters, large payload carriers, and medevac ambulances (Mars Institute 2020).

5.1.3.1.2 Habitability Studies, Experiences and Lessons Learned

Research at HMP is divided into two programs: Science and Exploration. The Science program is centered on studies of the ecological and biological biome of the Haughton Crater as a possible comparative model for the evolution of the Moon, Mars and other planetary bases. The Exploration program uses the site to develop, test and validate new exploration technologies and strategies involving habitats, spacesuits, ground vehicles, aircraft—drones and other unmanned aerial vehicles (UAVs), robotic rovers, drills, instruments, tools, life support systems, plant growth systems, human factors, crew management and communications and other information systems. Due to the fact that the base is only accessible and utilized during the short arctic summer, those visiting HMP are typically there for very short durations and since they all live in tents, studies regarding the habitability of the living space of the base is very minimal. However, factors such as the Arctic day and night cycle and restricted logistics and communications capabilities offer relevant conditions to serve as an analogue for the challenges of long-duration space flights and human and robotic exploration of the Moon and Mars (Mars Institute 2020).

5.1.3.2 Flashline Mars Arctic Research Station (FMARS)

> Our crews at FMARS are required to conduct a sustained program of geological, microbiological and climatological field exploration in a cold and dangerous remote environment while operating under many of the same constraints that a human crew would face on Mars. It is only under these conditions, where the crew is trying hard to get real scientific work done, while dealing with bulky equipment, cold, danger, discomfort, as well as isolation, that the real stresses of a human Mars mission can be encountered, and the methods for dealing with them mastered. It is only under these conditions that all sorts of problems that Mars explorers will face can be driven into the open so they can be dealt with. Only by doing these missions can we make ourselves ready to go to Mars. Nothing like this has ever been done before. [Dr. Robert Zubrin, President, The Mars Society (Flashline Mars Arctic Research Station, 2017)]

FMARS is the first Mars Analogue Research habitat built and organized by the Mars Society[6] with the specific intent to focus on human performance and adaptation to an ICE environment. The research habitat is in the polar desert of Devon Island, on Haynes Ridge, next to the Haughton impact crater, a 23 km (14 mi) diameter crater, seen as a Mars analogue environment due to its harsh conditions. The facility supports six-person crews for typically two- to eight-week seasons (see Fig. 5.13a, b).

5.1.3.2.1 Facility Architecture and Design

The habitat is a two-story 8.8 m (26.3 ft) diameter and 7.7 m (25.13 ft) tall cylindrical structure. It was first occupied in July and August, 2000 (Clancey 2000). The lower

[6] More information can be obtained via: http://fmars.marssociety.org/about-the-fmars/.

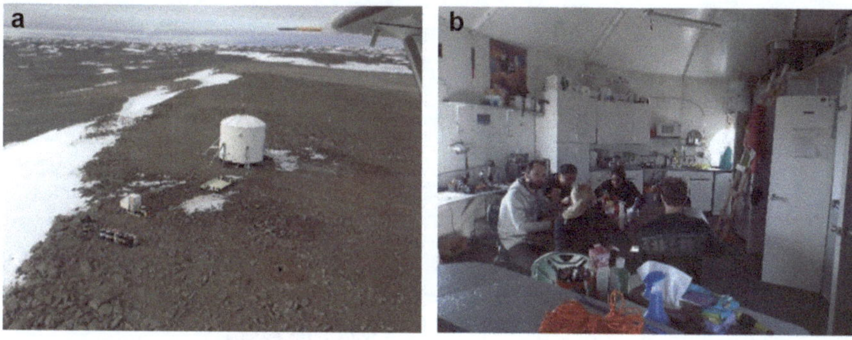

Fig. 5.13 (**a**) FMARS image outside (Credit: FMars, Mars Society, courtesy of Keith Keplinger). (**b**) FMARS Mars 160 crew sharing a meal (Credit: FMars, Mars Society) FMARS Mars 160 Crew Members: Alexandre Mangeot, Yusuke Murakami, Jonathan Clarke, Anastasiya Stepanova, Anushree Srivastava and Paul Knightly

deck is used for entry and exit and hosts the EVA preparation room with a toilet and a personal hygiene space. The upper deck hosts the crew quarters and the social area, including the galley, wardroom and food storage (see Fig. 5.14). A ladder connects both floors. The wardroom table is used for having meals together and also for work meetings and computers. The galley includes a stove, microwave and water container. The crew quarters have staggered bunkbeds. According to chief scientist William J. Clancey (2000) rooms are not equal in volume. Some are a bit broader than others. Water is brought in from a river a few hundred meters down the hill. Electricity is produced by a gas generator.

> FMARS is mostly a wooden monocoque structure, composed of double skinned wooden panels, except for the roof. [. . .] the main weight is carried by steel members driven into the ground, stabilised by the legs. FMARS also has steel guy ropes attacked to large pegs. As a result, FMARS feels much more solid, especially in high winds. Internally the doors on FMARS are square with a high lip... On the upper deck the position of the galley and stairs are reversed in FMARS compared to MDRS, the upper deck of FMARS retains the original configuration ... The lower deck of FMARS has more small compartments (including a separate tool room), and the toilet and bathroom are reversed. Jonathan Clarke, Mars 160 participant in both phases at MDRS and FMARS, 2021.

5.1.3.2.2 Habitability Studies, Experiences and Lessons Learned

From 2001 to 2005, ten crews rotated through FMARS for short duration missions, typically 2–4 weeks, with crews of 6–7 members. In the summer of 2007, the first 4-month-long FMARS mission was successfully completed with a crew of seven, Crew 11, and a full complement of research studies covering technology, human factors, medicine, psychology, and communications. Results from this first attempt to extend, heretofore, short duration missions in a habitat developed specifically for research (instead of operational environments such as bases in Antarctica) revealed a number of issues (e.g., multicultural conflicts, coping inefficiencies) that

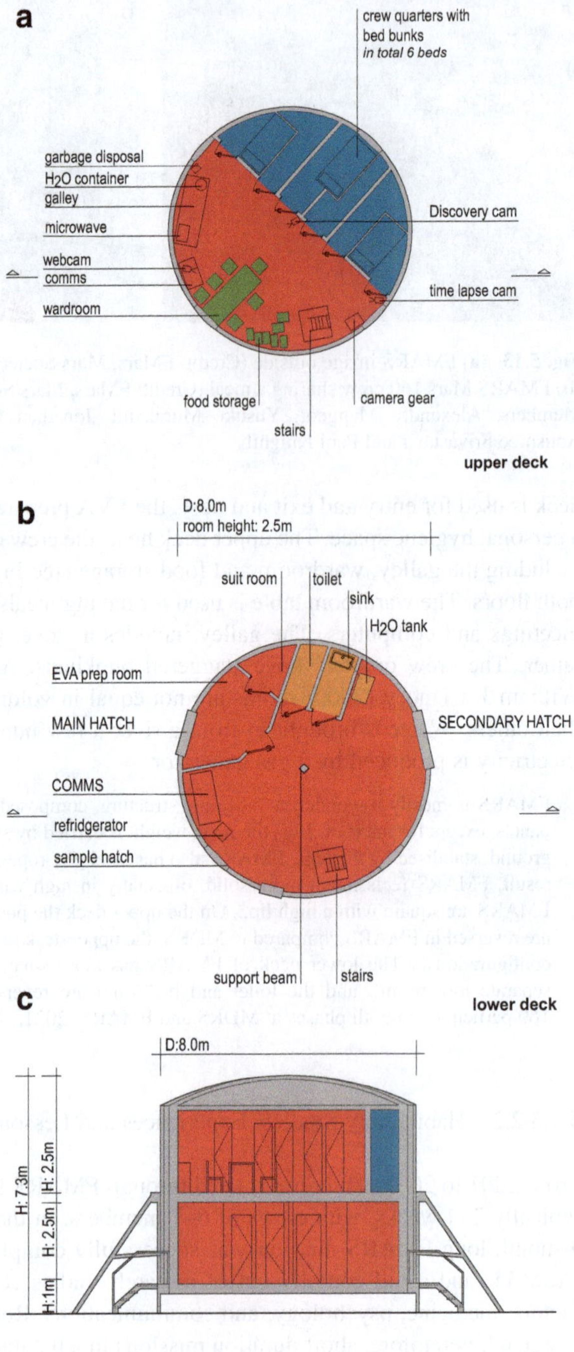

Fig. 5.14 Schematic plans (**a**,**b**) and section (**c**) of Flashline Mars Arctic Research Station Upper and Lower Decks (Credit: The Authors)

underscored the need for longer studies situated in real environments that pose true isolation and risk (Bishop et al. 2010).

A return to short duration missions was resumed in 2009 and 2013. In 2017, a crew of six arrived to complete the artic portion of a long duration mission originally intended to be 365 days (Mars 365) but which lack of funding had evolved into two shorter long duration missions (Mars 160). Conducted in two phases, the crew first rotated at a sister habitat, the Mars Desert Research Station (MDRS), in the desert of Utah, USA (80 days), and then followed with a month-long rotation at FMARS. The program was intended to provide a 'twin study' wherein the same crew would conduct the same research at both field stations.

5.1.3.3 Mars Desert Research Station (MDRS)

MDRS, is the second analogue research facility built and maintained by the Mars Society (see Figs. 5.15a and 5.16). It came operational in December 2001 and is situated in the Utah desert in the American Southwest, near Hanksville. MDRS' mandate has been to educate researchers, students and the public about the critical factors needed to settle Mars. The station is typically used by small crews for durations of 1 week to 3 months. The ongoing missions can frequently be viewed by the public via integrated webcams in the habitat.

As of April, 2020, 19 field seasons hosting 236 crews of 6–7 people have participated in missions typically lasting 1–2 weeks. Because of its ease of access, the American station is considered well suited as a test bed for equipment that will later be sent to more remote and unforgiving locations. For the same reason, the station has been the focus of short-duration isolation and confinement studies since its inception.

A wide range of psychological studies investigating crew factors in short-duration missions has been in place since 2002 (Bishop et al. 2004, 2006). Several longer duration missions have also been conducted at MDRS, most notably the first

Fig. 5.15 (**a**) MDRS image outside. (**b**) MDRS Crewmembers prepare for lunch inside The Mars Society's Mars Desert Research Station in Utah (Credit: The Mars Society, courtesy of Robert Zubrin)

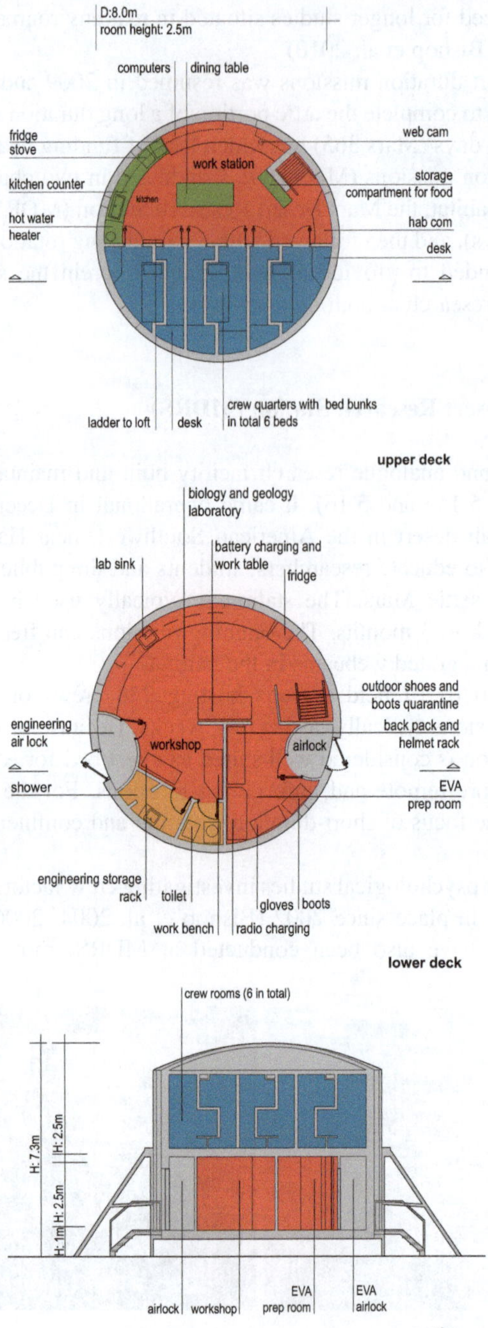

Fig. 5.16 Schematic plan of MDRS horizontal and vertical sections (Credit: The Authors)

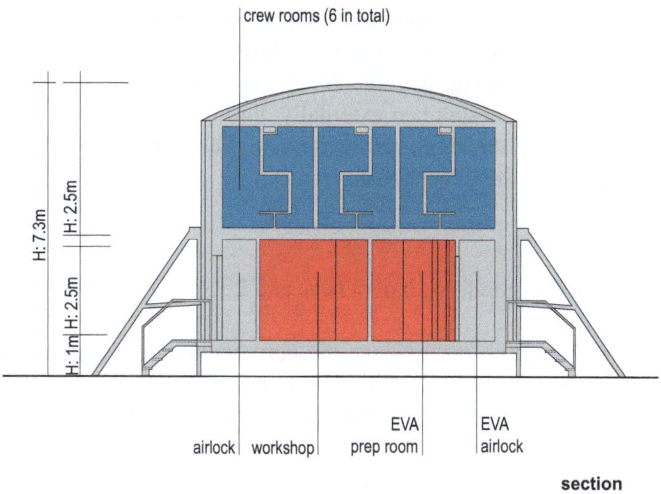

Fig. 5.16 (continued)

segment of the Mars 160 mission conducted in 2017 with 80 days at MDRS in which the same crew rotated through two different analogues (MDRS and FMARS) with similar structural designs but different extreme environments (Box 5.3). MDRS's crew roster has included national and international teams, university sponsored student educational rotations, NASA/ESA sponsored field teams, survival training teams, technology field tests and research studies investigating the entire gamut of human factors, engineering, medicine, psycho-social and habitat issues.

5.1.3.3.1 Facility Architecture and Design

The facility includes a two-story main building (the Hab), a greenhouse (the GreenHab), a solar observatory (the Musk Observatory), a robotic observatory, and engineering pod (the RAM) and a science building (Science Dome). All buildings, except for the robotic observatory, have been connected via tunnels since 2017 (see Fig. 5.15a). The Hab is a two-story cylindrical structure with a diameter of 10 m (33 ft) (see Fig. 5.16). The lower deck is dedicated to science and engineering work activities. There are two airlocks and the EVA preparation room, as well as storage and maintenance for space suits and their associated equipment. There is also a biology and geology laboratory, as well as the main hygiene area, including a shower and a toilet.

The upper deck is the social and private area. Here are six private crew quarters with small bunk beds and a small desk. There is an additional crew quarter in the loft area for a seventh crew member if needed. The open plan social space includes a dining table, work and entertaining area, a dedicated communication station, and a kitchen. The small kitchen is equipped with a gas stove, refrigerator, microwave oven, and a sink. The internal freshwater tank is stored under the dome shaped roof in the loft area. Here is also a maintenance access hatch for maintaining the antenna and other weather monitoring instruments. Power is supplied by batteries, located under the habitat. Water is supplied externally, through a water tank that is located 30 m (100 ft) away in the Engineering Support Equipment Area. For the toilet, greywater is used from the Greenhouse.

5.1.3.3.2 Habitability Studies, Experiences and Lessons Learned

Research has been a core activity since the very beginning. Each team member is expected to conduct at least one primary study during their rotation to provide meaningful work activity, a component shown to be of critical importance for individual and group wellbeing in ICEs. The research proposals are reviewed by a multi-national panel of experts for scientific soundness, feasibility and adherence to standards of ethical research. This means that over 1400 studies have been conducted at MDRS, compiling perhaps the most extensive track record across analogues. Comparisons between homogeneous-gendered teams, between mission teams and backup crews, and international cultural factors have been successfully conducted at MDRS, among others (Bishop et al. 2005, 2006; The Mars Society 2020).

However, human habitability studies on the impact of the living space as experienced by its inhabitants have been few as the short duration of most missions and small sample sizes of crews (n ~ 6) have placed greater emphasis on individual-level adaptation to confinement and group dynamics. For instance, a study proposed to investigate environmental noise at MDRS was, in fact, investigating the efficacy of using music and sounds of nature as a countermeasure to psychological stress (Ono et al. 2013) rather than the impact of the noise of the habitat on human performance and wellbeing. This is the challenge of all short duration facilities—there simply is not enough duration exposure to environmental and/or facility characteristics to provide a meaningful evaluation of impact. Nevertheless, MDRS has amassed an impressive history of analogue opportunity to an unmatched spectrum of scientists, educators, engineers, and human factors researchers.

Box 5.3 The same station design in two different environments

The Mars 160 Twin Analog Mission (M160) was a two-phase simulated Mars mission that took place over 140 days between 2016–2017 at the Mars Society's two analogue facilities—the Mars Desert Research Station (MDRS) near Hanksville, Utah (80 days) and the Flashline Mars Arctic Research Station (FMARS) on Devon Island, Canada (60 days). Originally built from the same basic base design, MDRS has evolved to a greater degree over the years due to its continuous use and more central location which

(continued)

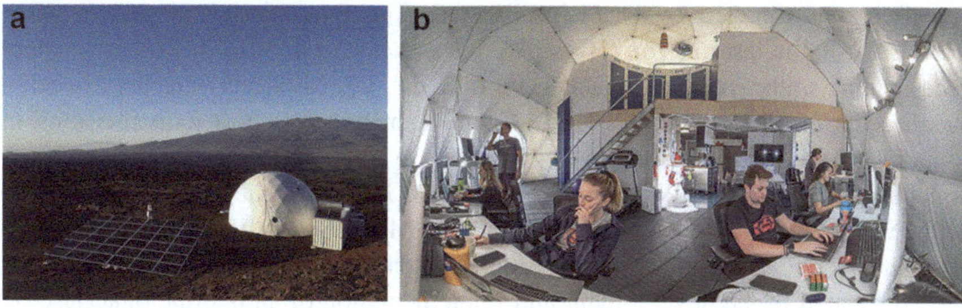

Fig. 5.17 (a) The HI-SEAS habitat on Mauna Loa, Hawaii (Credit: HI-SEAS, Michaela Musilova). (b) The HI-SEAS habitat, view from the interior (Credit: HI-SEAS, Epic Tours)

Box 5.3 (continued)

facilitates repairs and renovations. FMARS, on the other hand, presents significant logistic challenges for renovations and maintenance and required additional stabilization against high winds and thicker walls to protect against temperature extremes in the Arctic. Jonathan Clarke was a geologist that participated in both phases of the Mars 160 mission and reflected on some of the notable difference between the two habitats.

> FMARs was of much higher standard structurally than MDRS but in a much worse state, due to the more severely weather and greater difficulties and less frequent maintenance. Power was much more restricted at FMARS, less of it, and available for only a few hours a day, as noted. This did impact on your battery on your equipment. Lots of things were broken or substandard, causing much frustration. There were areas of rotting wood and (again) mold. There was lots of rubbish and broken equipment left over from the last, making the interior much more cluttered.
>
> FMARS is of course much more isolated than MDRS. For me this manifested itself in two ways. First, because of the limited internet bandwidth, there was far fewer demands from mission support, which was a great relief. We could just get on with the job. Plus, the sense of isolation being up in the Arctic was amazing, I really liked it. A bit like what Mike Collins said about orbiting over the far side of the Moon, not only not having mission control yakking in your ear, but also a sense of exultation. [Jonathan Clarke, personal communication, January 2021.]

5.1.3.4 HI-SEAS

The HI-SEAS[7] habitat (see Fig. 5.17a) is located in a remote quarry on the northern slope of Mauna Loa, Hawaii. From 2013 to 2018 the University of Hawaii at Manoa ran long duration Mars exploration analogue studies with funding from NASA, lasting 4 to 12 months. The program, which will be called **HI-SEAS Phase 1**

[7]Hawaii Space Exploration Analog and Simulation.

hereafter, was directed by Dr. Kim Binsted. The first mission started in 2013 (HI-SEAS 1), the last mission (HI-SEAS VI) was halted in 2018 due to an accident necessitating a crewmember evacuation. In 2018, Dr. Michaela Musilova became the director of **Phase 2** of **HI-SEAS** and began hosting short-term missions in cooperation with the EuroMoonMars Initiative, a program founded by the International Lunar Exploration Working Group (ILEWG 2019), a public forum sponsored by the world's space agencies, and International Moon Base Alliance, an association of scientists, educators and entrepreneurs from space agencies and industries worldwide dedicated to the development and implementation of an international base on the moon (International Moonbase Alliance 2019).

5.1.3.4.1 Facility Architecture and Design

The geodesic habitation dome (see Fig. 5.17b) has an internal two-story open layout with a diameter of about 11 m (36 ft). The ground floor is comprised of the communal areas with the kitchen, dining room, common workspace and lab, an exercise area and a small bathroom with a shower and toilet (cf. Häuplik-Meusburger 2017).

The ground floor (see Fig. 5.18) has an area of 30.3 m^2 (993 ft^2) (usable 26.8 m^2/878 ft^2) with the second floor spanning an area of 39.4 m^2 (424 ft^2) comprising six separate bedrooms and a toilet with a washing basin. Attached to the exterior of the habitat is a single converted shipping container (48.8 m^2/160 ft^2), providing storage for food and other supplies as well as hosting the water and electrical systems.

5.1.3.4.2 Habitability Studies, Experiences and Lessons Learned

In the **HISEAS-Phase 1 simulation missions**, behavioral research studies focused on "*the need to identify psychological and psycho-social factors, measures, and combinations that can be used to compose and support highly effective teams for future self-directing long duration exploration missions*" (cf. Häuplik-Meusburger 2014).

During the first 4-month simulation mission (HI-SEAS I), diverse food preparation strategies for long-term space exploration were tested by the crewmembers. As a result, mission commander Angelo Vermeulen confirmed the use of spices and higher fiber foods as important for ICE habitation in which the selection of food is severely limited. The importance of food systems has been highlighted by early astronauts during space missions and led to great improvements in food preparation as well as packaging. The use of spices and the individual mixing ('cooking') of ingredients allows a lot of variation and adds to "renewed pleasure" (cf. Häuplik-Meusburger 2011, p. 218).

The second long-duration simulation mission, HI-SEAS II, was 8 months long. Data was collected on team cohesion and performance with various methods and technologies, such as sociometric badges and 3D Virtual reality applications.

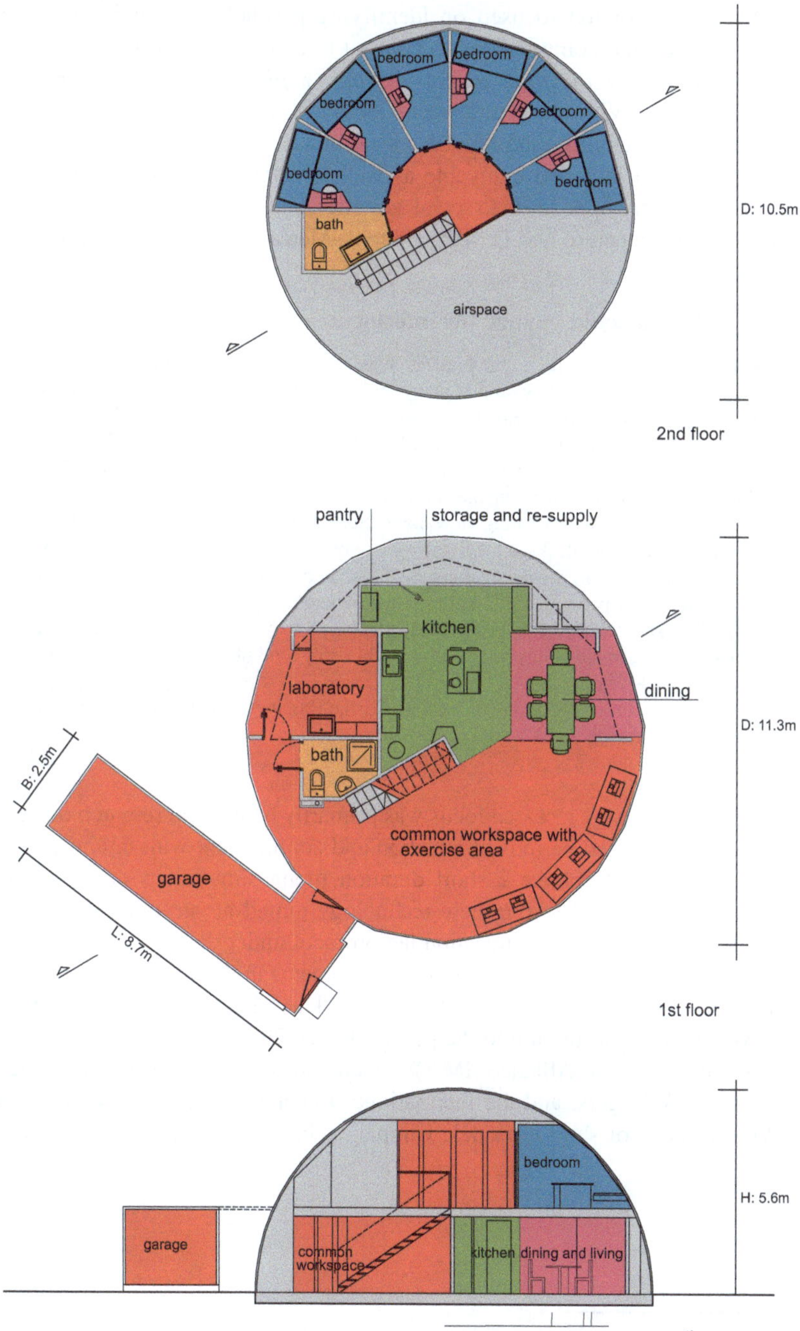

Fig. 5.18 Schematic plan of HI-SEAS horizontal and vertical sections (Credit: The Authors)

HI-SEAS III (8 months) focused on identifying psychological and psychosocial factors, measures and combinations to be used to compose "*highly effective crews for autonomous long duration and/or distance exploration mission*" (HI-SEAS [MediaKit] 2016). With the beginning of the 4th 8-month mission, HI-SEAS IV, the HI-SEAS Habitability study had been introduced. Selected socio-spatial phenomena of interest included enjoyable as well as unpleasant social activities and situations, privacy as opposition to social activities, and relationships and personality in relation to Architecture and Design (Häuplik-Meusburger 2014) (Box 5.4).

Box 5.4 If you could change the interior …

As far as the architecture … the biggest changes – I mean, we definitely would've needed some soundproofing so that you could actually go into your room and close the door and feel like you were alone instead of always hearing the treadmill, always having the events going into your room, letting in the smells from the kitchen. You can hear somebody talking in a normal voice almost anywhere in the dome. And then, while the stairs going straight up the middle of the dome to that landing, it's nice. You could go up there and have a big view down. Sort of social. At the same time, if you are at any of the work desks in the living room, you can see the kitchen and the dining room, into the airlock, private quarters. See if anybody's door is open or closed. The dome is almost a panopticon,[8] and you know where everybody is all the time, and you always know that you are being a little bit observed by everybody. So sort of lack of private space or the ability to get away could really kind of wear on you. So if you could change – yeah!

Source: Excerpt from an Interview with Tristan Bassingthwaighte, HI-SEAS long duration crew member, 2008 (not published)

Whereas the **Phase 1** research focus was primarily behavioral research on human adaptation to prolonged periods of isolation and confinement with minimal contact with Earth, **HI-SEAS Phase 2** short duration moon simulations are focused on science experiments and testing new technology aimed at surviving on the lunar surface. The 20-minute communication lag protocol under the Mars mission profile has been shortened to 3 seconds and installed cameras in the common spaces of the habitat are being considered to be left on 24/7. There have even been discussions about live-streaming a mission to the public (Koren 2018). In cooperation with the International Moonbase Alliance (IMA), which had been founded by the owner of HI-SEAS, Henk Rogers, and the International Lunar Exploration Working Group (ILEWG), a series of short-term field campaigns have been run since 2018 (Foing 2019).

[8]The panopticon is a type of institutional building and a system of control designed by the English philosopher and social theorist Jeremy Bentham in the eighteenth century. The concept of the design is to allow all prisoners of an institution to be observed by a single security guard, without the inmates being able to tell whether they are being watched.

5.2 Habitability Studies in Extraterrestrial In-Situ Environments

The examples of this group are all experiences from the real environment, the extraterrestrial environment. Although there has not been a habitat simulator in an extraterrestrial environment, the experiences of past space stations are considered as important analogues for a future space station. Outer space is the only environment in which the activities of the involved humans are embedded in the full-scale environment (e.g., microgravity, radiation) that will characterize future long duration habitats. However, to date, all existing extraterrestrial bases and vessels have been in low Earth orbit which provides some capability for rescue, contact and support that Lunar or Mars bases will not provide. So, in that respect, they are still 'analogues' for the bases of the future.

With reference to the Technology Levels used at NASA, this group encompasses the TRL Levels 7–9:

- TRL 7—System prototype demonstration in a space environment
- TRL 8—Actual system completed and "flight qualified" through test and demonstration (ground or space)
- TRL 9—Actual system "flight proven" through successful mission operations

5.2.1 *Salyut Space Station*

The Soviet Salyut program [russ. Salute, Firework] can be considered the first space station program. On 19 April, 1971, the Soviet Union launched its first space station, Salyut 1. Subsequently, the Soviets launched a series of single-module space stations in order to compete with the US in the space race. The program consisted of military (Almaz) and civilian space stations (Salyut). Salyut 7 was the last space station of this program and was in orbit from 1982 for almost 9 years (cf. Häuplik-Meusburger 2011).

5.2.1.1 Facility Architecture and Design

The Salyut space stations were designed as single-module 'Long-Duration-Stations' [russ. Acronym DOS-type station] whose architecture was constantly improved upon (see Fig. 5.19). Salyut 6 and 7 (see Fig. 5.20a, b), the second generation of the Salyut program had two docking ports to allow visiting crews to dock. Salyut 7 was the last and most comfortable space station of the Salyut series. Living conditions were constantly improved as each additional station was deployed, and with every station, modifications to the interior made it more habitable. For instance, there were approximately 20 windows with shades on the Salyut 7 (Häuplik-

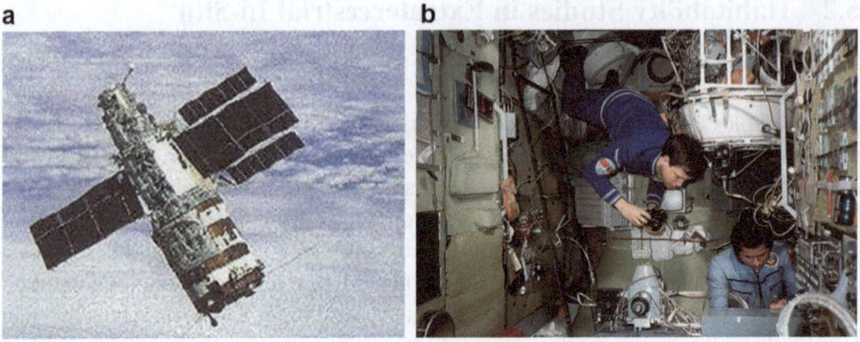

Fig. 5.19 (**a**) The Salyut-7 space station as seen by the Soyuz T-13 crew, Vladimir Dzhanibekov, Viktor Savinykh and Georgi Grechko. (**b**) The Salyut space station—interior. Cosmonaut Dorin Prunariu and Victor Savinykh on board of Saliut-6 space station, May 1981 (Credit: Dorin Prunariu)

Meusburger 2011, p. 47). Salyut stations (1971–1986) were laid out according to human activities, and had dedicated zones for working, sleeping, hygiene and exercise. Central was a table for work and having dinner together. From the beginning, the interior design was based upon a color scheme for supporting the cosmonauts' orientation in weightlessness, which was developed by Galina Balashova (see also Fig. 3.2b in Chap. 3).

5.2.1.2 Habitability Studies, Experiences and Lessons Learned

The stations were designed for long-term habitation and a vast amount of habitability studies were applied. Little was known on how people would actually live in a microgravity environment for long durations (Box 5.5). Aside from biological and psychological research, gaining knowledge on the production of food was considered vital. Since a pioneering OASIS greenhouse on Salyut 1 it had been demonstrated that plants could be grown in microgravity. In later Salyut stations many different models of plant growth facilities were tested. The cosmonauts planted onions, tomatoes, cucumbers and lettuce. The first space-grown vegetables were reportedly eaten in 1975 onboard Salyut 4 (Häuplik-Meusburger 2014). However, the Malachite greenhouse had planted orchids and was only used because of expected psychological benefits for the crew. While on his record breaking 211 days on-board Salyut 7, Valentin Lebedev wrote in his diary that for him plants were *like pets* (Lebedev 1990). He even placed his sleeping bag next to the OASIS greenhouse in order to look at the plants before going to sleep (Zimmerman 2003).

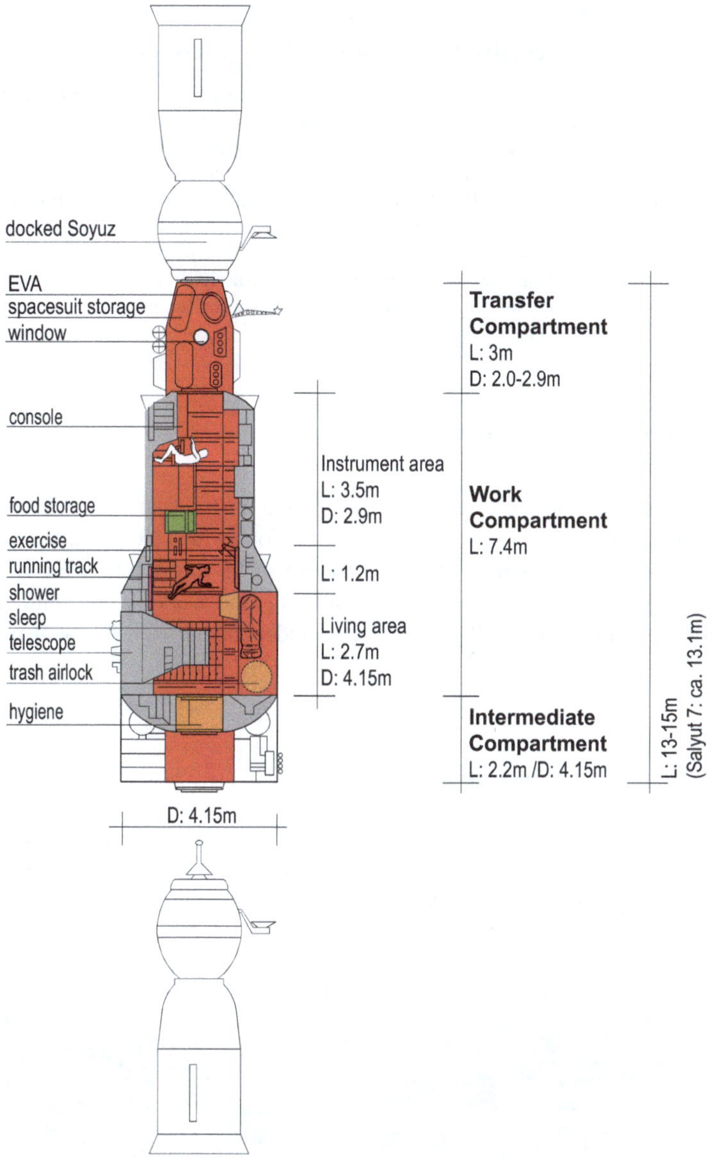

Fig. 5.20 Salyut 7 schematic diagram (Credit: The Authors, adapted from Häuplik-Meusburger (2011))

Box 5.5 Space Is Perceived Differently in Weightlessness

We docked […] The hatch finally opened. […] The first thought that crossed my mind when we entered the station was that this was going to be our home for the months to come. I was surprised that I didn't recognize it, maybe because I came in at an unusual angle—I walked in on the side panels. Weightlessness makes everything different. […] When we studied the station on the ground, our whole perception of it came from knowing what was up and what was down, where the floor was and where the ceiling was. But in space that doesn't matter. In the same space you can see many different interiors, depending on the position of your body. In other words, it's as though you could imagine several different rooms from one furnished room.

Source: As published in the book Lebedev 1990, p. 26.

5.2.2 *Skylab Space Station*

Skylab was the first space station for the United States (see Fig. 5.21a). It was launched on 14 April, 1973, and hosted three missions with a crew of three men each (Skylab 2: 28 days; Skylab 3: 59 days; Skylab 4: 83 days). The overall mission objective of Skylab was to create and test a living and working environment for a long period of time that is self-sufficient and independent from frequent re-supply (see Fig. 5.21b).

5.2.2.1 Facility Architecture and Design

Skylab was the largest space station in diameter (6.25 m) ever built, as its architecture was formed by a converted third stage of a Saturn V Moon rocket (see Fig. 5.22). The former Saturn V hydrogen tank was converted into a 14 m

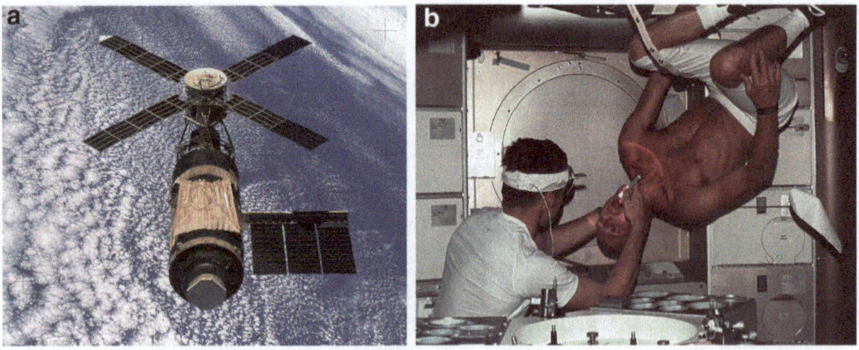

Fig. 5.21 (**a**) The Skylab space station—exterior, as photographed from the Skylab 4 Command and Service Module (CSM) during the final fly-around before returning home (Credit: NASA). (**b**) The Skylab space station—interior. A medical checkup aboard Skylab. Skylab 1 commander Pete Conrad undergoes a dental examination by medical officer Joseph Kerwin (Credit: NASA)

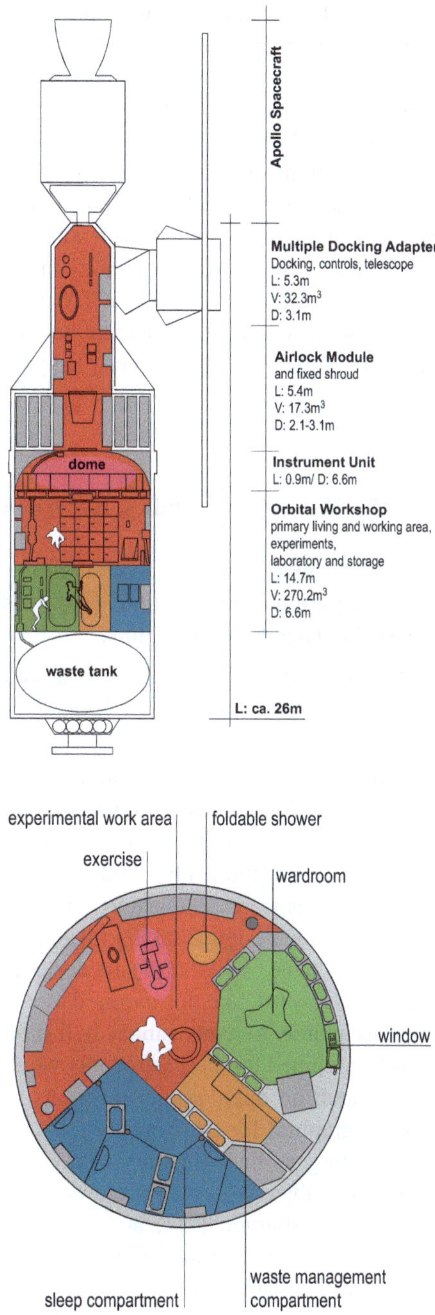

Fig. 5.22 Schematic plan and section of the Skylab space station (Credit: the Authors, adapted from Häuplik-Meusburger (2011))

two-story living and working space, which was named the Orbital Workshop Module (Fig. 5.21a).

The Skylab station had many dedicated architectural features enhancing its habitability. It had a dedicated area for food preparation and dining, with a triangular table designed by the industrial designer Raymond Loewy to avoid creating hierarchical positions for crew members during long missions. While having dinner together, astronauts could look out of a big window (see Box 3.1 in Chap. 3). The Skylab station was equipped with well-designed hygiene equipment and provided a deployable shower, which was a cylindrical foldable device situated in the experiment and work area in the Orbital Workshop (OWS). However, each Skylab crew seemed to have a different reaction to the shower. The first Skylab crew (Skylab 2) liked the shower. The second crew (Skylab 3) did not like the shower at all (Häuplik-Meusburger 2011).

5.2.2.2 Habitability Studies, Experiences and Lessons Learned

The station was designed and configured with the best understanding of the habitability requirements at that time. Post mission experiences were well documented in the Skylab Experience Bulletins.

> The biomedical program encompassed six areas of study: general clinical evaluations, neurophysiology, musculoskeletal functions, body fluid biochemistry and hematology, cardiovascular function, and nutrition and metabolic function (NASA [Skylab] 2004).

A wide range of habitability studies and experiments were conducted under the supervision of the industrial designer Raymond Loewy and NASA. Skylab's long duration missions and large open volume provided the opportunity to—for the first time—document the neutral body position in space. On Skylab missions different body restraints were demonstrated and evaluated, and it became clear that workstations had to be designed taking microgravity conditions into account. For instance, the dining table, so laboriously designed by Lowey, couldn't be used properly because it was flat, and this required astronauts to uncomfortably tense their stomach muscles in order to remain above it. It should have been designed at a sloped angle so that the slumped forward astronauts would have been parallel to it.

Research on microgravity positions led to the well-known 'microgravity body posture graphic' for the Man-Systems Integration Standard (1985). However, the data leading to that graphic is based only on the measurements of 3 (!) male American subjects (Skylab 4 crew) and 12 photos (NASA [Bull.17] 1975, p. 13; Häuplik-Meusburger 2011, pp. 18–19). Today, we know that the range between those positions vary much more and are also related to the individual astronaut. However, there are some similarities that have large consequences for the design of the whole space station interior: **The head is tilted down, the arms and legs lift up, and there is no pooling of fluids in the lower extremities** (see Chap. 7, Fig. 7.2b).

5.2.3 *Mir Space Station*

The Mir Space Station was launched in 1986 and had 14 years of lifetime in orbit (see Fig. 5.23a, b). The stations hosted 105 cosmonauts from 11 nationalities, and thus became the first international space station. Mir was originally planned for long duration missions hosting two crew members. After 10 years, the Shuttle-Mir program (1994–1998) began, and later up to 6 cosmonauts stayed on-board. The Mir space station was constantly occupied from 1989 to 1999. In 1994 Valeri Polykov spent 438 continuous days on the space station.

5.2.3.1 Facility Architecture and Design

The Mir space station was the first truly modular space station due to its six docking ports (see Fig. 5.24). All modules were launched unmanned and docked automatically to the core module. In total, the Mir space station complex encompassed about 380 m^3. The core module was about 90 m^3 with a maximum diameter of 4.35 m. Its six modules encompassed the astrophysics module Kvant 1, the additional living module Kvant 2 with an airlock, the Earth observation module Kristall, the geophysical module Spektr and the ecology research module Priroda.

Overall, the architecture was similar to the Salyut space stations. The Core Module (Base Block) was the main living module with two private crew cabins (with windows) and used colors to mimic 'Earth orientation' (cf. Häuplik-Meusburger 2011, p. 72).

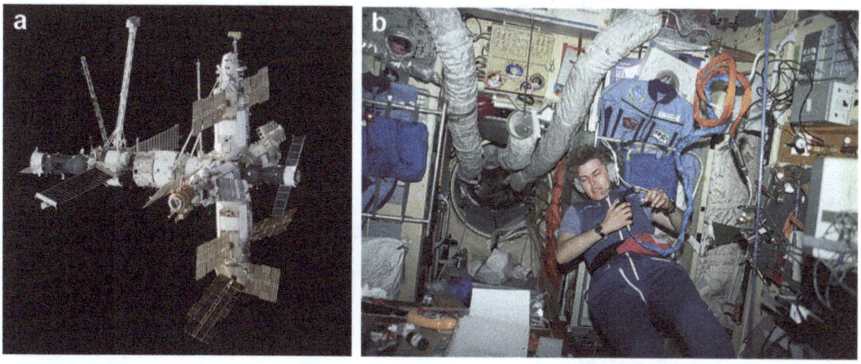

Fig. 5.23 (**a**) Approach view of the Mir Space Station viewed from Space Shuttle Endeavour during the STS-89 rendezvous. A Progress cargo ship is attached on the left, a Soyuz human operated spacecraft attached on the right (Credit: NASA). (**b**) Astronaut Shannon W. Lucid communicates with the ground support team inside the Core Module of the Russian Mir Space Station. Launched aboard the STS-76, the third Shuttle/Mir docking mission, in March 1996, to join the Mir crew in the orbiting laboratory, Astronaut Lucid returned to Earth aboard STS-79 in September 1996. At the time, Lucid made the U.S. longest record of 188 days in space (Credit: NASA)

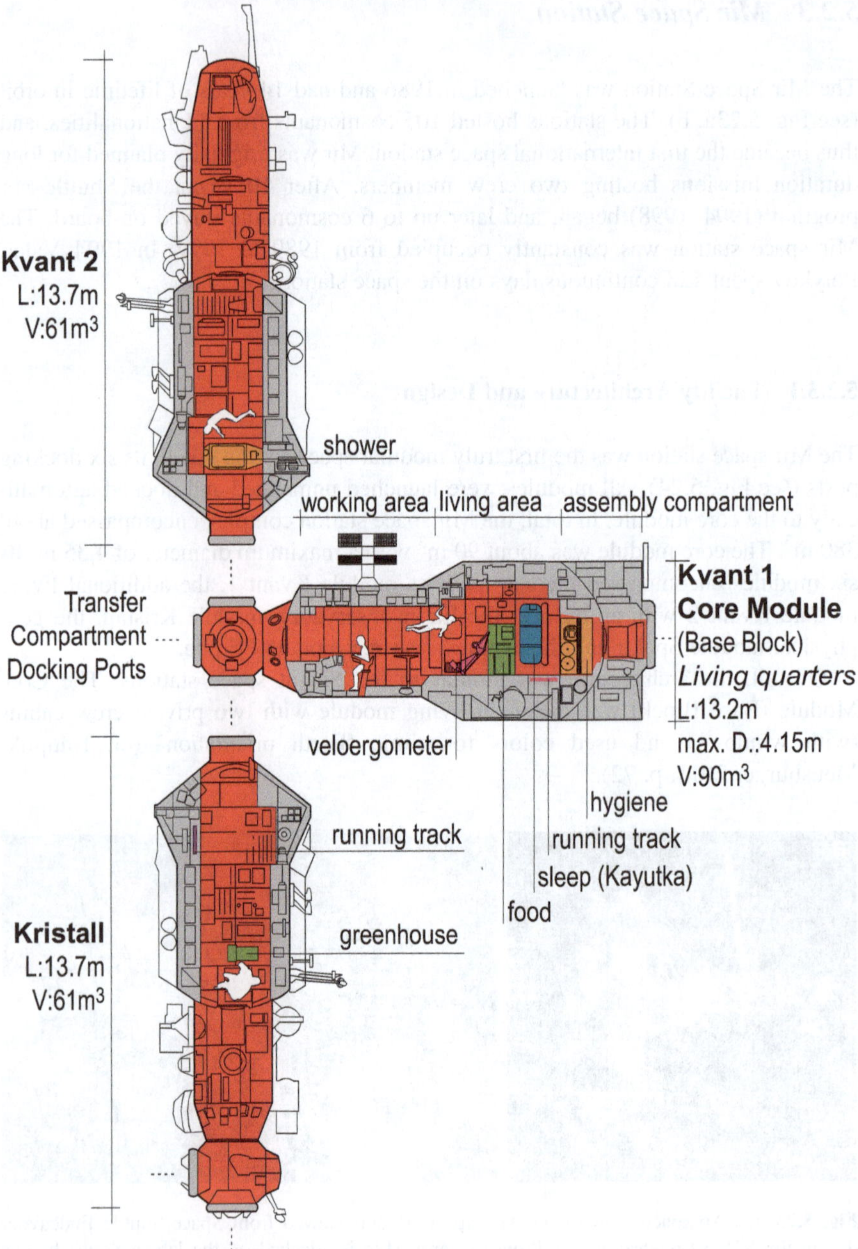

Fig. 5.24 Schematic plan of Mir space station (Credit: The Authors, adapted from Häuplik-Meusburger (2011))

5.2.3.2 Habitability Studies, Experiences and Lessons Learned

The experiences of human adaption to living and working on-board Mir have been very important for designing the International Space Station. Many of the findings over Mir's long history relating to human adaptation to long duration stays in the station have been qualitative as the 'sample size' has typically been 1–3 (at most) crew members at any one time. However, it was the first space station inhabited by an international crew and for very long periods of time. In the later years, when Mir was exceeding its lifetime, and was occupied by more people than it was designed for, life-threating situations increased (Box 5.6).

Throughout the years, indications that the impoverished environment of the station generated maladaptive responses has been repeatedly documented. For instance, it was noted that rage was a common emotional expression in early Mir crews which was later attributed to the sensory-poor environment and inadequate ability to communicate (Vessel and Russo 2015). There were also incidences of flight missions in the 1970s and 1980s being terminated early due to psychological factors (Cooper 1976).

Box 5.6 Lessons on Crew Autonomy

One reason for the cosmonaut's autonomy despite scheduled work might be that Salyut and Mir cosmonauts had no constant communication with ground at the time. They could only communicate via line-of-sight circuits. As the Soviets could use only the stations on their home territory and on ships at sea, their spacecraft were out of radio contact for at least one-third of each orbit, and often more. During the fire on Mir in 1997, the crew was out of communication range with mission control in Moscow for about half an hour.

The ham radio would have worked, but that meant "*broadcasting to the blind*" (Portree 1995, p. 106). At this time "*only 20 minutes were available for two-way communication per 90-minute orbit*" (Foale 1999, p. 96). The cosmonauts had to make a decision without mission control in this situation. They decided not to leave the station and fought the fire successfully.

Text passage, as *published in the book Häuplik-Meusburger 2011, in Chapter 7.7 Summary Observation, p. 250.*

The Mir-22 crew included Valeri Korzun (commander), Alexander Kaleri, Vasily Tsibliev, Aleksandr Lazutkin, Reinhold Ewald, and Jerry Lineger.

5.2.4 Spacelab Module

The Spacelab module was also amongst the earlier of extraterrestrial modules developed that integrated habitability; it was flown on 16 Space Shuttle missions

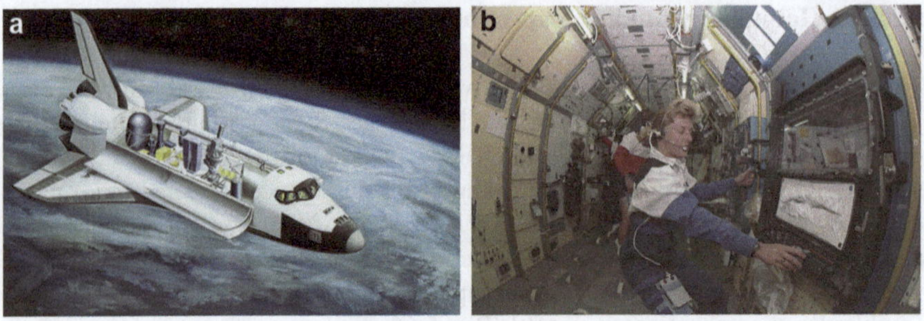

Fig. 5.25 (**a**) This illustration depicts the configuration of the Spacelab-2 in the cargo bay of the orbiter (Credit: NASA). (**b**) Astronaut Jan Davis worked inside the Spacelab-J module in space shuttle Endeavour during STS-47 in 1992 (Credit: NASA)

in the 1980s and 1990s.[9] Although the focus on 'habitability' was at the most fundamental level, the Habitable module expanded the volume for astronauts to work in a shirt-sleeve environment and had space for equipment racks and related support equipment. The scientific laboratory module was jointly developed by ESA and NASA and was designed and produced by ESA as "*an integral part of the Space Shuttle system*" (Rumerman 2009, p. 224).

5.2.4.1 Facility Architecture and Design

The Spacelab module was a cylindrical reusable laboratory module designed to fit in the Shuttle's cargo bay (see Fig. 5.25a). As such it had an outer diameter of 4.12 m (13.5 ft). It consisted of a pressurized laboratory module and non-pressurized pallets. The module could be configured in multiple ways, with either of one or two segments, each segment having a length of 2.7 m (8.9 ft) (see Fig. 5.26). When both segments were used, the maximum outside length was 7 m (23 ft) (Rumerman 2009).

5.2.4.2 Habitability Studies, Experiences and Lessons Learned

Life Sciences studies were executed during STS-40 (1991), STS-73 (1995) and STS-78 (1996), microgravity studies during STS-55 and STS-78 (1993).

> Crews worked around the clock in two shifts to staff this unique laboratory so that the maximum amount of science could be achieved. Spacelab-1 proved many experiments could be performed in multiple disciplines at the same time—a concept still used by station crews today (NASA [Spacelab] 2013) (see Fig. 5.25b).

[9]The first mission (Spacelab 1) with LM1 was flown with STS-9 Columbia in 1983.

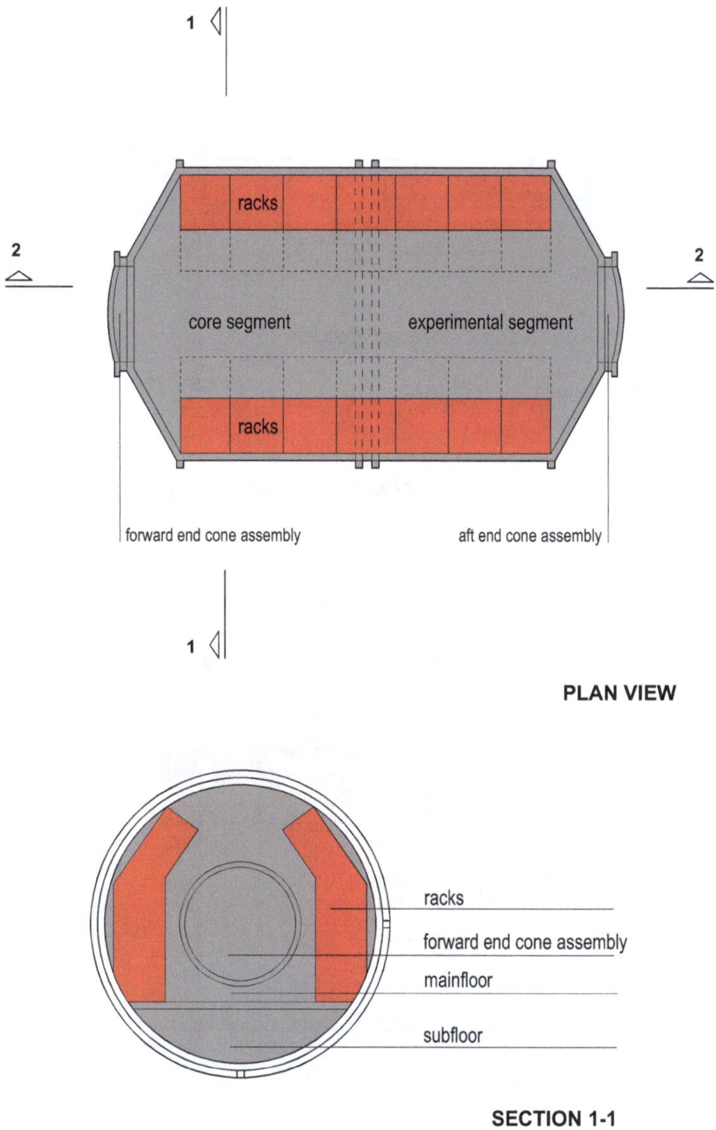

Fig. 5.26 Schematic plan of the Spacelab (Credit: The Authors)

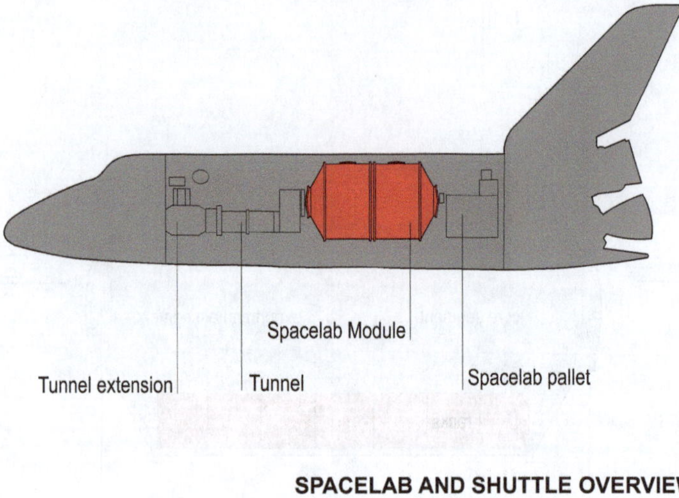

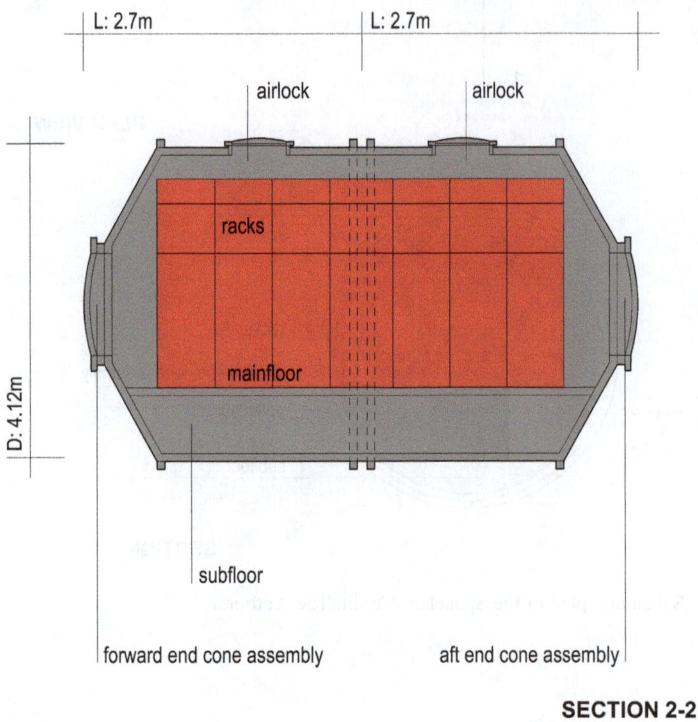

Fig. 5.26 (continued)

The experience and lessons learned of the Spacelab missions also helped design the International space station. For example, a multiple-user rack system has been first tested during the STS-94 mission in 1997. Today the International Standard Payload Rack (ISPR) is used as reconfigurable payload accommodation in the US, Japanese, and European modules. A number of design and procedure challenges were also identified. For instance, the connecting tunnel between modules had a sharp right-angle bend in it. This disoriented astronauts every time they went through it and came out in a different orientation than when they entered. Foot restraints at various workstations underwent numerous revisions as various improvements to ground-based designs were made to accommodate the reality of microgravity body postures and maneuverability. The U.S. requirement mandating English Only transmissions between Earth and SpaceLab also proved to be very challenging for international participants as these modifications were developed.

5.2.5 *The International Space Station (ISS)*

The International Space Station is the ninth space station to be inhabited by crew (see Fig. 5.27a). It is a compilation of modular units including living spaces in low Earth orbit built by a multi-national collaboration between five space agencies: NASA (United States), Roscomos (Russia), JAXA (Japan), ESA (Europe) and CSA (Canada). The facility serves as a microgravity and space environment research laboratory conducting research into astrobiology, astronomy, meteorology, physics, space medicine, life sciences, and behavioral research on human adaptation to the ICE environment of space (see Fig. 5.27b).

The station has been continuously occupied since November 2000 but had crews in residence even during the 42 assembly flights. Crews have typically ranged from 2 to 6 members from member organizations with Peggy Whitson setting the record for the most total time living and working in space at 665 days (NASA 2020).

Fig. 5.27 (**a**) ISS exterior (Credit: NASA). (**b**) Roscomos Flight Engineers (from left) Anatoly Ivnishin and Iven Vagner practice a simulated emergency evacuation of the ISS (Credit: NASA)

5.2.5.1 Facility Architecture and Design

The facility is 108.8 m (357 ft) long with 916 m^3 (32,333 ft^3) pressurized and 377.83 m^3 (13,696 ft^3) of habitable volume, not including visiting vehicles of which six spaceships can be connected to the ISS at the same time (see Fig. 5.28). The habitable volume available (divided into a "Russian Orbital Segment (ROS)" and an "U.S. Orbital Segment (USOS)") consists of 14 main pressurized modules plus lifeboats and other small additional volumes such as visiting cargo vehicles.

Although one big facility, architecturally the Russian and the US segment are very different. In principle, the US and European interior is dominated by the reconfigurable standardized rack configuration with a (more or less) free aisle and a functional space around, with the interior based on the cylindrical envelope to be transported with the Shuttle. The interior of the Russian segment is an updated continuation of the experiences from the Salyut and Mir Architecture. Also, color schemes vary.

There are, for example, two crew quarters on the Russian side and four on the US side, which look quite different. The two crew quarters in the Zvezda module, were provided early on (2000), are fully equipped and have a porthole window. For 8 years those were the only permanent sleeping places, while the other astronauts were sleeping in various other places in the station in sleeping bags. The first four rack-sized crew quarters were only installed in 2008. They are now circularly arranged in the Harmony Node 2, and provide private, and soundproof booths with good ventilation.[10]

Building on experience and technology development from earlier space stations, the ISS currently has two Water Recovery Systems—the one in Zvezda processes water vapor from the atmosphere. It could be consumed in an emergency but is typically fed to the Elektron system to produce oxygen. The new system in the American segment was installed in 2008 that can process water vapor collected from the atmosphere and urine into water that is intended for drinking (Carter 2009).

5.2.5.2 Habitability Studies, Experiences and Lessons Learned

As with many habitability improvements, a gap must be discovered first in practice. With a permanent habitation for over 20 years, these are a few examples. A poignant case in point was the issue of locking devices being put into place on the outward opening hatch on the Space Shuttle. Transcripts from the NASA Johnson Space Center Oral History Project indicate that impromptu locks were used by crews on several shuttle missions before they were permanently put into place (Kelley et al. 2016). From incidents involving failed experiments triggering implied threats of suicide (Reichhardt 2002) or simply discomfort with unfamiliar crewmembers,

[10]Good ventilation is a serious concern since a bubble of exhaled carbon dioxide can form around the head of sleeping astronauts causing oxygen deprivation while sleeping.

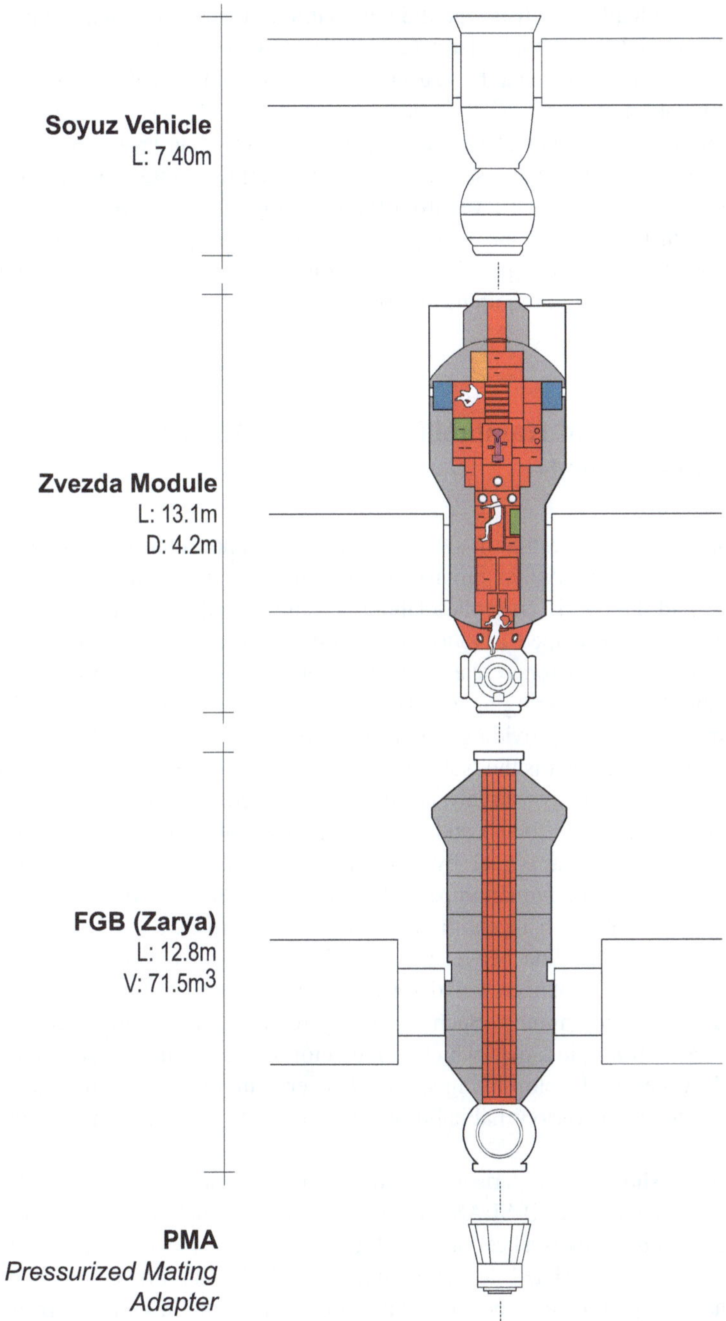

Fig. 5.28 Schematic plan of ISS (Credit: The Authors, adapted from Häuplik-Meusburger 2011)

especially payload specialists who did not go through the same extensive training as full astronauts (Shaw 2002; Hartsfield 2001; O'Connor 2006), the decision to padlock the hatch was made by astronauts to prevent any accidental or intentional depressurization of the vessel often enough to add these devices as permanent fixes.

Similar to other earlier space stations, noise was often an issue and was directly associated with sleep problems. Next to the use of earplugs, music was used to cover the high background noise level from the life support system and the ventilators. Since the installation of the first prototypes for a sleep station, acoustic noise mitigation, along with integrated radiation reduction, controllable airflow and warnings systems are design requirements for crew quarters.

5.3 Summary of Habitability Studies in In-Situ Environments

Overall, the use of terrestrial in-situ analogue environments has provided the bridge between purely operational environments with all the extraneous uncontrolled elements and the overly controlled laboratory and capsule habitats (see Table 5.1). As such they have helped advance habitability research from humans-in-the-loop studies (i.e., human factors research) to evaluations of human response to the habitat itself. This shift has far-ranging effects. For instance, human factors studies wherein the human is simply providing assessments of some aspect of the environment (where the environment is the 'subject' of focus; e.g., how appealing, how comfortable, how spacious) have largely not been considered human research (where the human response is the focus) and, therefore, have not been routinely required to undergo review and approval by Institutional Review Boards (IRB) or Ethics Committees. Once the human impact of the environment upon the human becomes the focus, such research meets the criteria as human research and such review is mandatory. This fine line between human factors/engineering design and human subject research has not been well defined nor widely appreciated in the architectural and engineering community. Such regulatory review is frequently experienced as intrusive and stifling in a field which has, heretofore, been able to design with a free hand. However, to develop living spaces that serve human needs, the research into human responses to various habitability factors must be the next central focus in that journey.

Table 5.1 shows a comparison of some of the terrestrial in-situ habitat characteristics. For instance, the HI-SEAS Facility has a dome structure, and the underwater facilities feature cylindrical ceilings, both of which present different ceiling heights. HERA, MDRS and FMARS feature cylindrical walls but uniform ceiling heights yet different in shape for each floor (i.e., flat ceilings on the first floor but domed ceiling on the second floor). Characteristics such as these not only present challenges to furnishing spaces but also represent different perceptions and experiences of the space.

Table 5.1 Summary and comparison of terrestrial in-situ habitat features

IN-SITU TERRESTRIAL ENVIRONMENTS: Habitat features					
Name of facility	Module type	Environment	Total area	Simulation duration	Features
TEKTITE I & II	2 vertical cylinders, 2 stories, connected by tunnel	Underwater	~48 m^2 + tunnel	10, 20, 60 days	Total dependency on life support; experiments focus on ecology, biology, decompression sickness, microbiology, mycology, isolation, confinement
AQUARIUS/ NEEMO	Horizontal cylinder, one-story	Underwater	41 m^2	7, 14 days	Total dependency on life support; experiments focus on isolation, confinement, communications, telemedicine, and remote health care technologies
AMUNDSEN-SCOTT SOUTH POLE STATION	Modular 2-story, adjustable, elevated	Antarctica	7400 m^2	9 m	Bioregenerative life support technology and recycling systems, human health and performance, adaptation to extreme isolation and confinement
VOSTOK STATION	5 aluminum frame buildings	Antarctica	Unknown	12 m	Ice core drilling and magnetometry, actinometry, geophysics, medicine and climatology
CONCORDIA	2 elevated towers with 3 floors each	Antarctica	1500 m^2	9 m	Studies psychosocial responses to extended isolation and confinement, crew cohesion, task performance
FMARS	2 story cylindrical	Arctic	8.81 m diameter	2–4 w, 1 and 4 m	Studies psychosocial responses to extended isolation and confinement, crew cohesion, task performance and food study

(continued)

Table 5.1 (continued)

IN-SITU TERRESTRIAL ENVIRONMENTS: Habitat features					
Name of facility	Module type	Environment	Total area	Simulation duration	Features
MDRS	2 story cylindrical	Desert	10 m diameter	1–2 w, 80 d	Studies psycho-social responses to extended isolation and confinement, crew cohesion, task performance, use of VR in maintaining social connections, and food study
HI-SEAS	Dome, two story interior structure; attached shipping container for storage	Volcano	145.8 m^2/ 1462 ft^2 usable	4, 8, and 12 m	Studies psycho-social responses to extended isolation and confinement, crew cohesion, task performance, use of VR in maintaining social connections, and food study

Note: Module Type, total area and simulations duration of Tektite, Aquarius, Concordia, FMARS, MDRS and HISEAS missions are listed from Häuplik-Meusburger (2014). HMP not listed due to the extensive use of tents for crew quarters

Table 5.2 shows a comparison of some of the extraterrestrial in-situ habitat characteristics. For instance the early space stations are all single element modules. Mir was the first modular space station and with the ISS the first inflatable module was tested. The spatial orientation is mainly horizontal and work oriented. The orientation has only been different for non-working activities, such as sleeping, hygiene and exercise.

Once it became apparent that habitability was a critical factor in wellbeing in ICE environments, attention to extraterrestrial in-situ habitability factors became the focus. In an extensive 2016 NASA review of risk for adverse cognitive or behavioral conditions and psychiatric disorders, Kelley et al. (2016) combined data from both terrestrial analogues as well as extraterrestrial flights regarding probable sources of stress and disfunction for future long duration crews. The impact of isolation, confinement and adjustment to both terrestrial and the unusual environment of space reaffirmed the risk of emotional, behavioral and cognitive impacts found in terrestrial analogues. The report noted that the same types of countermeasures used in ground-based ICE environments were effective for orbital crews as well, e.g., staying connected with family/friends through electronic media, providing a variety of leisure activities and the importance of food. They also noted that long duration, more remote bases on the Moon and Mars would pose clear challenges for these

Table 5.2 In-situ extraterrestrial facilities: Habitation volume, duration and crew of historical space habitats and the International Space Station. Sources: Cohen (2009) and Häuplik-Meusburger (2011)

Extraterrestrial facilities: Habitat features					
Name of facility	Module type	Spatial orientation	Habitable volume [m^3]	Mission duration	Features
Apollo CM (with LM) Ended 1972	Single elements; Command Module, Service Module, Service Module, Lunar Module, Lunar Roving Vehicle	CM: horizontal and vertical LM: horizontal	10 (6)	12 d 12 m (Apollo 15); On Lunar Surface: 3 d 2 h 59 m (Apollo 17)	11 missions, 6 lunar landings; almost constant communication (line of sight) with delay
Salyut 7 Ended 1984	Single element space station with docking possibilities	Horizontal; exceptions: exercise equipment, windows; use of colour code	90	16–237	One resident crew/2–3) and visiting crews (2–3); space station series with gradual change of hardware
Skylab Ended 1979	Single element space station with multiple docking adapter	Mainly horizontal: upper deck: free space; Exceptions: toilet	320	28, 59, and 84	Three crews (3CM); used once; converted rocket; testbed for a number of restraints and interior design features
Mir Ended 2001	Modular space station; Automated launch and configuration	Similar to Salyut stations	380 (core module: 90)		First international space station, with an original design life of five years, inhabited 1986–2000, constantly occupied from 1989–1999, Valeri Polyakov spent 438 days at once; Limited communication with mission control
Spacelab Ended 1998	Single module; science extension to fit the Shuttle's cargo bay	Horizontal	75 (4.1×7)	8–10 d Shuttle missions with 12 h shifts	Was flown on 16 Space Shuttle missions; reusable laboratory module; multi-configurable
International Space Station (ISS)	Modular space station with truss-backbone structure; Shuttle and robotic arm needed for configuration in orbit	Mainly horizontal	1200	Varies, 180–340;	Inhabited since 2000, since 2009 by a permanent crew of six; Recycling of air, wastewater, urine and production of oxygen; Expandable concept with inflatable modules; constant communication and monitoring via telemetry

traditional approaches to be effectively employed as distance would interject longer time periods between resupply, restrictions in provision of new resources and communication time lags.

References

Bamsey, M.T., Zabel, P., Zeidler, C., Gyimesi, D., Schubert, D., Kohlberg, E., Mendedoht, D., Rae, J., & Graham, T. (2015), Review of Antarctic greenhouses and plant production facilities: A historical account of food plants on the ice. In *45th International Conference on Environmental Systems*, 12–16 July, Bellevue, Washington.

Barrett, E., & Martin, P. (2014). *Extreme: Why some people thrive at the limits*. Oxford: OUP.

Bennett, J. (2016). *How Antarctic isolation affects the mind. Canadian psychologist Peter Suedfeld is studying stress in the world's harshest environment*. Retrieved June 18, 2020, from https://www.canadiangeographic.ca/article/how-antarctic-isolation-affects-mind

Bishop, S. L. (2011). From earth analogs to space: Getting there from here. In D. Vakoch (Ed.), *Psychology of space exploration: Contemporary research in historical perspective* (1st ed., pp. 47–77). Washington, DC: NASA SP-2011-4411.

Bishop, S. L., Dawson, S., Rawat, N., Reynolds, K., & Eggins, R. (2004). Expedition one: Assessing group dynamics in a desert Mars simulation. In *55th International Astronautical Congress*, Vancouver, 4–7 October 2004.

Bishop, S. L., Sundaresan, A., Pacros, A., Patricio, R., & Annes, R. (2005). A comparison of homogeneous male and female teams in a Mars simulation. In *56th International Astronautical Congress*, Fukuoka, Japan, October 2005.

Bishop, S. L., Dawson, S., Rawat, N., Reynolds, K., Eggins, R., & Bunzelek, K. (2006). Assessing teams in Mars simulation habitats: Lessons learned from 2002-2004. Mars Analog Research. *American Astronautical Society Science and Technology Series, 111*, 177–196.

Bishop, S. L., Kobrick, R., Battler, M., & Binsted, K. (2010). FMARS 2007: Stress and coping in an arctic Mars simulation. *Acta Astronautica, 66*(9–10), 1353–1367.

Candidi, M., & Lori, A. (2003). Status of the Antarctic Base at Dome C and perspectives for astrophysics. *Memorie della Societa Astronomica Italiana, 74*(29), 29–36.

Carter, D. L. (2009). *Status of the regenerative ECLSS water recovery system*. Retrieved July 28, 2020, from https://ntrs.nasa.gov/archive/nasa/casi.ntrs.nasa.gov/20090033097.pdf

Clancey, W. J. (2000). *A framework for analog studies of Mars surface operations: Using the Flashline Mars Arctic Research Station*. Retrieved May 18, 2020, from https://ti.arc.nasa.gov/m/pub-archive/210h/0210%20(Clancey).pdf

Cool Antarctica, Antarctic Stations – Bases – Currently Occupied. Retrieved from https://www.coolantarctica.com/Community/antarctic_bases.php

Cooper, H. S. F., Jr. (1976). *A house in space*. New York: Holt, Rinehart and Winston.

European Space Agency (ESA). (2005). *Mission to Mars via Antarctica*. Retrieved June 18, 2020, from http://www.esa.int/esaHS/SEMBZA8A9HE_research_0.html#subhead1

European Space Agency (ESA). (2013). *The remotest base on Earth*. Retrieved June 18, 2020, from http://www.esa.int/Science_Exploration/Human_and_Robotic_Exploration/Concordia/The_remotest_base_on_Earth

Ferraro Choi and Associates Ltd. (2017). *Elevated station design for the South Pole Redevelopment Project at Amundsen-Scott South Pole Station*. Retrieved June 18, 2020, from https://ferrarochoi.com/publications/elevated-station-design/

Flashline Mars Arctic Research Station. (2017). *Flashline Mars Arctic Research*. Retrieved July 1, 2020, from http://fmars.marssociety.org

Foale, C. (1999). *Waystation to the Stars*. London: Headline.

Foing, B. H. (2019). EuroMoonMars instruments, research, field campaigns, and activities 2017-2019. *LPI, 2132*, 3090.

Geoscience News and Information. (2020). *The world's largest deserts. A map showing the generalized location of Earth's ten largest deserts and a table of over 20 major deserts.* Retrieved July 27, 2020, from https://geology.com/records/largest-desert.shtml

Hartsfield, Jr. H. W. (2001). *Oral history transcript*. Interviewed by C. Butler. *NASA Johnson Space Center Oral History Project,* 6-12-2001. Retrieved May 21, 2020, from https://historycollection.jsc.nasa.gov/JSCHistoryPortal/history/oral_histories/HartsfieldHW/HartsfieldHW_6-12-01.htm

Häuplik-Meusburger, S. (2011). *Architecture for Astronauts: An activity-based approach.* Wien: Springer Science & Business Media.

Häuplik-Meusburger, S. (2014). Astronauts orbiting on their stomachs: The need to design for the consumption and production of food in space. *Journal of Applied Statistics, 84*(6), 114–117.

Häuplik-Meusburger, S., Binsted, K., & Bassingthwaighte, T. (2017). Habitability studies and full scale simulators: Preliminary themes following HISEAS mission IV. In *47th International Conference on Environmental Systems*, Charleston, South Carolina, 16–20 July 2017.

HI-SEAS [MediaKit]. (2016). *HI-SEAS Hawaii space exploration analog and simulation, Media Kit.* Retrieved 24 August 2020, from http://hi-seas.org/wp-content/uploads/2016/05/HISEASMediaKit_01APR2016.pdf

International Moonbase Alliance. (2019). *The International Moonbase Alliance*. Retrieved June 16, 2020, from https://moonbasealliance.com/

International Moonbase Alliance (ILEWG). (2019). International Lunar Exploration Working Group (ILEWG). Retrieved May 8, 2020, from https://moonbasealliance.com/ilewg (ILEWG 2020).

Jarvis, O. (2016). *Project Tektite: The Aquanauts that lived in the sea*. Retrieved May 18, 2020, from https://www.uw360.asia/project-tektite-the-aquanauts-that-lived-in-the-sea/

Joyner, C. C., & Chopra, S. K. (1988). *The Antarctic legal regime*. Dordrecht: Martinus Nijhoff.

Kelley, S. J., Williams, T. J., Schneiderman, J. S., Whitmire, A. M., Picano, J. J., Leveton, L. B., Schmidt. L. L., & Shea, C. (2016). *Evidence Report: Risk of adverse cognitive or behavioral conditions and psychiatric disorders*. Retrieved May 18, 2020, from https://ntrs.nasa.gov/archive/nasa/casi.ntrs.nasa.gov/20160004365.pdf

Koren, M. (2018). Hawaii's Mars simulations are turning into Moon missions: The dome where crew members practiced red-planet missions will now be converted to a simulated moon base. *The Atlantic*. Retrieved May 8, 2020, from https://www.theatlantic.com/science/archive/2018/11/hi-seas-mars-habitat-nasa-moon-simulation/576295/

Lebedev, V. (1990). *Diary of a cosmonaut – 211 days in space*. New York: Bantam Books.

NASA. (1972). *Tektite II habitability research program*. Retrieved June 1, 2020, from https://ntrs.nasa.gov/archive/nasa/casi.ntrs.nasa.gov/19720007419.pdf (NASA [Tektite II] 1972).

NASA. (2006). *About Aquarius*. Retrieved May 15, 2020, from https://www.nasa.gov/mission_pages/NEEMO/facilities.html (NASA [Aquarius] 2006).

NASA. (2013). *Spacelab paved critical path to Space Station*. Retrieved May 8, 2020, from https://www.nasa.gov/mission_pages/station/research/news/spacelab/ (NASA [Spacelab] 2013).

NASA. (2019). *Haughton Mars project (HMP)*. Retrieved May 25, 2020, from https://www.nasa.gov/analogs/hmp (NASA [HMP] 2019).

NASA. (2020). *International Space Station Facts and Figures*. Retrieved May 8, 2020, from https://www.nasa.gov/feature/facts-and-figures

NASA [Bull.17]. (1975). Skylab Experience Bulletin No. 17: Neutral Body Posture in Zero-G. Houston, Texas, USA: NASA, Lyndon B. Johnson Space Center. JSC-09551.

NASA Life Science Data Archive. (2004). *Skylab Program*. Retrieved July 1, 2020, https://lsda.jsc.nasa.gov/Research/research_detail/16?researchtype. (NASA [Skylab] 2004).

NASA, NEEMO-HABITABILTY. (2004). Habitability assessment: Habitat utilization analysis from human factors perspective (NEEMO-HABITABILITY), NEEMO 5. Retrieved 15 May 2020, from https://lsda.jsc.nasa.gov/Experiment/exper/1099

National Science Foundation (NSF). (2016). *NASA signs collaborative agreement to support human health research in polar environments*. Retrieved June 18, 2020, from http://www.spaceref.com/news/viewpr.html?pid=49474

Nicolas, M., Bishop, S. L., Weiss, K., & Gaudino, M. (2016). Social, occupational, and cultural adaptation during a 12-month wintering. *Antarctica, Aerospace Medicine and Human Performance, 87*, 1–9.

Nielsen, J., & Vollers, M. (2001). *Ice bound: A Doctor's incredible battle for survival at the south pole*. New York: Miramax Books.

O'Connor, B. D. (2006). Oral history transcript. Interviewed by S. Johnson. *NASA Johnson Space Center Oral History Project*. 4-20-2006. Retrieved May 21, 2020, from https://historycollection.jsc.nasa.gov/JSCHistoryPortal/history/oral_histories/OConnorBD/OConnorBD_4-20-06.htm

Ono, A., Schlacht, I. L., Hendrikse, J., & Battler, M. (2013). *Habitability in Mars mission simulation: Sounds as stress countermeasures*. In Lunar and Planetary Science Conference (Vol. 44), March 2013.

Pauli, D. C, & Cole, H. A. (1970). Summary report on Project Tektite I. A multiagency 60-day saturated dive conducted by the United States Navy, the National Aeronautics and Space Administration, the Department of the Interior, and the General Electric Company. Retrieved May 18, 2020, from https://ntrs.nasa.gov/archive/nasa/casi.ntrs.nasa.gov/19710016053.pdf

Portree, D. S. (1995). *Mir hardware heritage* (Vol. 1357). Houston, TX: Lyndon B. Johnson Space Center.

Reichard, S. E. (2010). *Who on earth is Sylvia Earle? Undersea explorer of the ocean*. Berkeley Heights: Enslow.

Reichhardt, T. (2002). *Space shuttle: The first 20 years (Air & Space)*. London: DK.

Rumerman, J. A. (2009). *Nasa historical data book* (Vol. VII). NASA launch systems, space transportation/human spaceflight, and space science 1989–1998. U.S. Government Printing Office.

Shaw, Jr. B.H. (2002). Oral history transcript. Interviewed by K. M. Rusnak. *NASA Johnson Space Center Oral History Project*. 4-19-2002. Accessed 21 May 2020, from http://www.jsc.nasa.gov/history/oral_histories/ShawBH/ShawBH_4-19-02.htm

Spaceflight Insider. (2013). *NASA's Tektite II Undersea Habitat: An interview with Aquanaut & Engineer Peggy Lucas Bond*. Retrieved June 12, 2020, from https://www.spaceflightinsider.com/space-flight-news/nasas-tektite-ii-undersea-habitat-an-interview-with-aquanaut-engineer-peggy-lucas-bond/

The Mars Institute. (2020). *Haughton-Mars Project (HMP)*. Retrieved June 14, 2020, from http://www.marsonearth.org/about/history.html

The Mars Society. (2020). *Our Mission*. Retrieved Retrieved August 24, 2020, from http://mdrs.marssociety.org/about/

Vessel, E. A., & Russo, S. (2015). *Effects of reduced sensory stimulation and assessment of countermeasures for sensory stimulation augmentation*. A report for NASA behavioral health and performance research: Sensory stimulation augmentation tools for long duration spaceflight. Retrieved May 8, 2020, from http://kinetic-cna.pl/web/pageFiles/kcfinder/files/adaptacja%20synaptyczna%20ARc.pdf

Weiss, K., Feliot-Rippeault, M., & Gaud, R. (2007). Uses of places and setting preferences in a French Antarctic station. *Environment and Behavior, 39*(2), 147–164.

Zimmerman, R. (2003). *Leaving earth: Space stations, rival superpowers, and the quest for interplanetary travel*. Washington: Joseph Henry Press.

Chapter 6
Projections

One cannot claim the importance of the lived experience in ICEs and then write an entire book without exploring the perspective of those who have been there. Thus, we intended to incorporate historical quotes and reflections from a broad gamut of those who have experienced ICE environments. Yet, we felt most of those reflections were incidental and, as such, may have missed important aspects from a more thoughtful consideration. So, while working on the book, we crafted a qualitative survey intended to prompt focused consideration on key aspects of habitability that seem to be important characteristics based on prior anecdotal reports and research observation and we sent it out to people who met the following qualifications:

- They had occupied a permanent structure (i.e., no tents, recreational vehicles or other temporary, transitional structures) for
- at least 2 weeks in duration located in
- a simulated or actual extreme environment that
- involved confinement and/or isolation.

We asked questions about the following aspects: (1) What was most problematical with living in the ICE? (2) What worked? (3) How will extraterrestrial habitats differ from our terrestrial habitats? and (4) How would incorporating desirable elements help adaptation to the extreme confinement and isolation conditions in a closed loop habitat? Short guest essays in Sect. 6.3 explore these questions.

6.1 Life in Extreme Environments

When contemplating the enormity of what is going to be entailed in the mundane details of surviving and thriving in an extraterrestrial location, the imagination stutters initially. Inhabitants will not only be more remotely separated from all known life more completely than in the entirety of human existence, but the very environment will entail myriad unknown challenges and dangers to be overcome.

S. Häuplik-Meusburger, S. Bishop, *Space Habitats and Habitability*, Space and Society, https://doi.org/10.1007/978-3-030-69740-2_6

The realization that one cannot envision what has yet to be experienced heightens the costs of failure. Space is an unforgiving environment to a greater magnitude than any we have yet attempted to populate. Past human diasporas have always operated within a relatively benign terrestrial envelope that provided basic environmental necessities (e.g., unlimited breathable atmosphere, access to new resources, gravity, ability to easily expand territory, mobility). None of these will be present once we leave Earth. All resources of any kind must be provided or accessed through technology and design. **So, how do we know what we don't know?**

The prior chapters (Chaps. 4 and 5) detailed human endeavor to systematically study various aspects of living in environments that simulate or replicate different factors of those future space habitats on our horizon. Most were terrestrially bounded and certainly lacked the capability to inform us about the effects of microgravity or high radiation or even the psychological impact of being cut-off from our terrestrial cradle. But they, and the small handful of orbital habitats to date, have contributed to our growing body of insight that will so critically help us proactively shape effective living spaces for extraterrestrial existence.

Scientific investigation is only one part of the pursuit of insight. Experiential knowledge is also required. Like so many other aspects of life, unless you have done it, it is only an intellectual exercise. True understanding is found embedded in the feelings, sensations, thoughts, emotions, and visceral responses to immersion within the *situation*. Ask any parent if the experience of the birth of their child was 'adequately' covered by all the classes, videos, shared stories of others … and they will simply laugh. Ask anyone who has successfully reached the summit, survived a dangerous mission, or outlasted an extended ordeal if they were 'prepared' for everything that happened, and they will tell you it was impossible to be prepared for everything. They simply had to manage in the moment. But, having done that, been there … they are better prepared for the unexpected.

> The missions were planned, but there was a lot of open time. We were doing a lot of different things that were not planned for (…), **because we didn't know what to plan for**. We had a lot of time for spontaneous observations and discovery.
>
> That's what exploration is all about. You have to be careful not to program too much (Harrison Schmitt, Apollo 17 Astronaut, 2009).[1]

So experiential knowledge is invaluable and only acquired by <u>*doing*</u>. This factor is so important that we wanted to dedicate an entire chapter to sharing feedback garnered from those that have spent time in various isolated, confined or extreme environments. We sent out an open-ended survey with semi-structured questions world-wide, shared through networks of colleagues, friends, professional groups, enthusiasts and content experts to those in the space community that have either participated in or been in a support role for those in ICE locations. We focused on experiences specifically related to living spaces that had some degree of permanence rather than transitory shelters such as tents. We asked for each respondent to identify

[1]Interview with Harrison Schmitt by Sandra Häuplik-Meusburger in Vienna, Austria, 2009. Transcript not published.

the kinds of ICE habitats they had experienced, their role (participant, observer or both) and respond to our questions from the perspective of the experience with the longest duration (as a participant if they had done both) and address factors that dealt with habitability.

Of course, we got so very much more than those focused reports. It's simply impossible to talk about the need for privacy (a major habitability design factor) without delving into the emotions and situations that surrounded the stress of constantly dealing with violations or invasions of privacy. Many of the comments were purely related to psychosocial factors of leadership, group roles, hardships generated by the environments themselves or administrative/organizational difficulties. On the surface, it would appear that a significant percentage of 'extraneous' responses were about anything but habitability! And yet...

Perhaps, there are features that can be incorporated into future habitats that will mitigate some of those 'extraneous' group dynamics, processes and procedures that contribute to effective group functioning. We believe so. Insight comes in many forms. Here is what our contributors have shared with us.

6.2 The Survey

The survey first asked for demographic details and ICE experience from the respondents to (1) ensure that they had the targeted experience (in-situ habitats located in simulated or real ICE environments) and (2) to provide us details about the kinds of individuals sharing their experiences. The second half asked about their experiences in the habitat with the longest duration. We then grouped the habitat experiences into three habitat types: (1) **Controlled/laboratory simulation HABs** (HERA, IBMP/Mars500, ICARUS, SpaceHaven), (2) **In-Situ HABs** (MDRS, HiSeas, FMARS, OEWF, Underwater habs, Biosphere; and Antarctica/Arctic, Research Vessels), and (3) **Space HABS** (Mir, ISS, Shuttle, Skylab). Group 1 corresponds to the historical review in Sect. 4.3. Group 2 correspond to the historical review in Sect. 5.1. And Group 3 corresponds to the historical review in Sect. 5.2. We thank all the contributors for sharing their insights. (See Preface).

6.2.1 The Participants

There were 45 respondents that met qualifications for the survey. Out of the 35 who gave personal information, the majority were males with a group average age of 44 years, representing 15 different countries (Figs. 6.1 and 6.2).

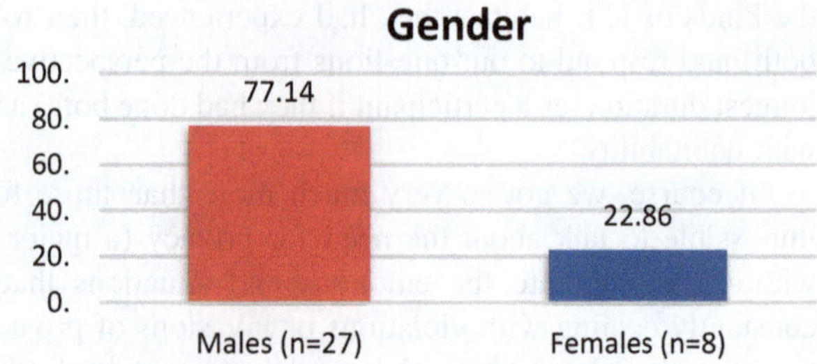

Fig. 6.1 Gender distribution

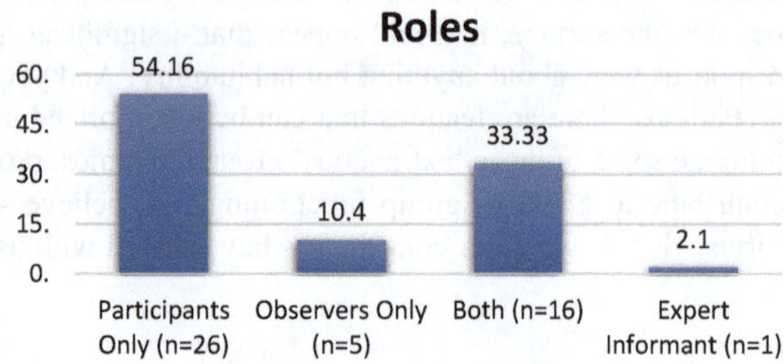

Fig. 6.2 Distribution across ICE roles

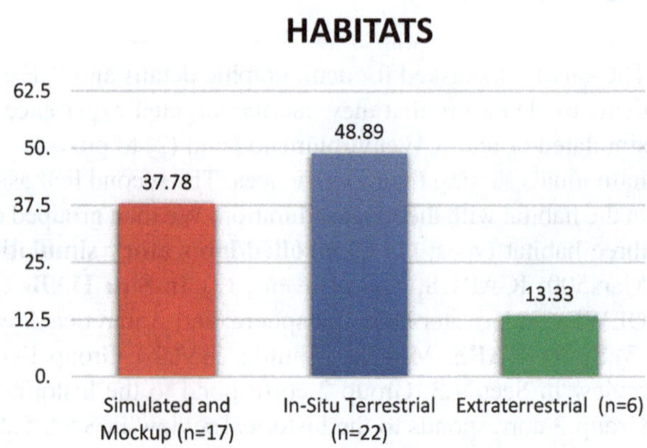

Fig. 6.3 Percentage of respondents with relevant experience in various ICEs

6.2.2 The ICEs

The largest majority were participants only but a substantial number (33%) had served as both participants and observers (Fig. 6.3).

Interestingly, 19 out of 45 participants (42.22%) spent time in two or more ICE habitations (range 1–4). The mean duration of stays was 93.15 days (sd = 153.86; median = 30; range = 7–730 days) when the extraordinary stay of 730 days (2 years) for a biospherian was included and 70.58 days (sd = 110.73, median = 30, range 7–395) without. There were several others with extensive durations of more than 300 + days from Antarctic bases, oceanic vessels, and HiSEAS.

6.3 Individual Reflections on Habitability by the Inhabitants

Below are participant responses from each question posed from the perspective of ICE inhabitants grouped by habitat type. Summary figures were crafted for those groups that had sufficient number of respondents to provide an adequate sampling for analyses. Word cloud (Wordclouds.co.uk) lists by necessity had to exclude many terms that were infrequently mentioned and similar terms (e.g., privacy and private spaces) were combined into the same categories. Quotes are just listed and are not commented or categorized beyond the habitat being referenced.

6.3.1 Experienced Challenges of the Living Space

(Q5) What physical, social, or psychological aspect(s) of the habitat living environment listed above caused you/or individuals/or the group the most discomfort?

6.3.1.1 Controlled/Laboratory Simulation HABs (Fig. 6.4)

Fig. 6.4 HERA Q5 Responses

[HERA] *The habitat didn't really cause much in the way of discomfort. The overall amount of space was adequate for work, sleep and hygiene. There were some minor ergonomic issues with table height vs. chair height that caused some strain during long work periods, but overall the mission schedule allowed for movement to different work stations throughout the day to minimize that discomfort. The crew were most often seated in close proximity and that we found to be acceptable, but occasional shifting of position was necessary due to long stretches of time with no tasks to complete.*

[HERA/MDRS] *Personally, for me, not been able to start the day with fresh air was the biggest discomfort. I was very excited about it, but the first day in the morning when I woke up with their alarm under white lights, that's when it really struck me as what I had signed up for. Overall, I had a positive experience, but it did make me realize how much I enjoy fresh air etc., and based on my experience at HERA, I have decided not to reapply for any similar isolation study. I really felt claustrophobic sitting in small planes after HERA [...]. MDRS was different, and I think I may redo it, if an opportunity arrives. To improve the situation of lack of fresh air, I am not sure what can be done, but I know some groups are exploring the use of virtual reality to create an illusion of being on a beach or nature, that may help.*

[HERA] *Regarding the habitat space/design, a suggestion that we all gave during debriefing was that a simple modification in the shower set up (the curtain rod) would be more practical: the shower area was very small, especially for tall/big people; having a curved curtain rod (instead of a straight one) would allow more mobility while showering and wouldn't have an impact on the usable space. Another suggestion we gave was to have desks that not only can be folded, but that can also be adjusted in high.*

[HERA] *Regarding habitat design, I thought HERA was amazingly efficient. The two design aspects that I recall being mildly annoying were: (1) on level 2, we were often in each other's way in the mornings and evenings when we needed to access our laptop computers. The computers were all stored in one location, and it was sometimes difficult to get to them quickly. (2) The lift from level 1 to level 2 was painfully slow.*

[HERA] *The biggest challenge for our HERA crew boiled down to interpersonal dynamics. [...] There were many days with no non-essential communication. I had never been in such a tense situation. Our crew did our best to address the issues, but it is clear we could have been equipped with better team management and conflict training. The few HERA activities that were designed to promote dialogue actually helped reduce tensions, and more of these activities earlier may have helped our situation. In terms of the physical aspects, I found the food and sleep situations to be perfectly comfortable. We were sleep deprived (5 hours per night), and it was difficult to tell if this added to our stress or not. [...] the once-per-week phone calls were huge for helping me feel connected to the family matters. I would have appreciated some more phone calls, although it's hard to say if I would have been more or less preoccupied by that if I was in closer contact. The 5-minute communication delay with MCC was interesting to manage. It was clearly a challenge for us, but at the same time we managed it well I think. From the perspective of a crewmember, when the comm delay increased, we were constantly debating what decisions we should make autonomously versus run by MCC. Going through MCC meant a time delay, so we began to operate more autonomously. [...] Basically, poor communication platforms were frustrating, and clear/fast communication was relieving.*

[HERA] *From a physical standpoint, there were occasional near misses (and hits) on my head walking through low hatchways and opening overhead storage compartments. The galley table ended up perpetually messy with papers, laptops, tablets, and books; I felt a little guilty using a laptop as an occasional place mat, but we were usually tidy with food; most of the time, we were eating directly from the pouches as we held them. Sometimes using the central elevator was helpful (example: porting the wastewater from the galley sink down to the bathroom for dumping; staging between floors to check the hydroponics experiment), while other times it was tedious (example: mission orders to forgo the ladder for the elevator exclusively during a period of intentional sleep deprivation, even if for safety purposes). There was plenty of storage, but even though many containers had labels, many did not... and for those that did, it was the classic indexing problem of "oh, that's the word I should have used" instead of the more colloquial term for something that might have come to mind (like "facemask" or "ventilator" instead of "PPE"). Labels are good, but sometimes a pictogram can go a long way. Some experiments required focus and concentration, forcing other crew members to stay very quiet and restrict their motion around certain parts of the hab. It was very easy to wander through an area making too much noise and distract or frustrate your crewmate if you weren't careful. This could be tricky, as many of these experiments were conducted in the first floor (lab area) of the core module, which was also an intersection with the airlock and—more importantly—the hygiene module (i.e., restroom). There was occasionally some second-guessing of words and actions, given the extensive arrays of microphones and video cameras throughout the hab; their presence was always in the back of my mind, and mission control would occasionally hop on the comms to correct us if they noticed something awry. There was a little bit of a "Big Brother is watching you" vibe... but then, we expected that going in.*

[HERA] *For myself (and my fellow crew mates would agree), I am not a morning person, in the sense that I like time by myself to wake up. During our mission, the sleep deprivation only made that fact abundantly clear. We spent the first couple of hours next to each other: donning our gear, eating breakfast, talking about how the newspaper hadn't arrived yet (when it should have), watching tv, etc. [...] Because I was more irritable in the morning, the crew member that I sat next would get the brunt of it. We realized the problem into the second week of the mission and it was made known to keep the socializing with me to a minimum for the first couple of hours. [...] Although we worked out the problem as best we could, we would have benefitted from more flexibility in the schedule during the beginning of the day.*

[HERA] *The lack of privacy was the most difficult part. We had instructions to stay in public areas except only when using the bathroom and sleeping. In real missions, such a mandate would not be enforced. Having the ability to "go private" after a while is pretty important in order to keep things from blowing up.*

[HERA] *[...] Another discomfort I had was lacking food. Although my resting metabolic rate was taken prior to Ingress, whatever algorithm NASA used to pack food rations is skewed. [...] Another discomfort I experienced was seeing a crewmember take a liking to another crewmember. Watching their flirting behavior and alone time in the bathroom (which was the only area that didn't have cameras), was uncomfortable.*

[HERA] *One of the challenges of an analogue study is keeping the crewmembers comfortable in a 1 g environment. The purpose of the analogue is to study group dynamics in a confined environment over a "long duration" mission. However, some of the data sensors that we wore routinely got in the way over the course of our day. Wrist watches, activity monitors, microphones to record conversations, and occasionally, a skin temperature sensor that would protrude from our foreheads for 36 hours. This was the most uncomfortable thing for me and made getting adequate rest a challenge. This kind of hardware wouldn't have been as cumbersome on the ISS, because we'd have been sleeping in microgravity, so where the sensor was in relation to your pillow wouldn't keep you up at night! For this issue, there really isn't a great solution that a habitat designer could incorporate. It falls mostly on the hardware that the investigator provides. [...] In all honesty, the module itself was very comfortable for me. I'd been a [military] Officer prior to my experience with HERA, so I'd grown accustomed to working in close quarters with other people for long periods of time. The crew quarters we had in HERA offered sufficient privacy and allowed for a kind of mental decompression period when we went to bed in the evening. Some people definitely are affected by lack of privacy after a prolonged period of time, but I didn't notice that in our group.*

[HERA] *Tribalism—the crew broke down into groups based on who was tasked to spend the most time together. This had a negative impact on overall psychological wellbeing as well as on our performance to accomplish tasks. Could be improved by ensuring that crew has strong relationships going into the mission and measures are taken during the mission to nip such dynamics in the bud before they become a problem. Training the crew to be aware of these types of psychological issues and how to deal with them might also be beneficial [...] Caffeine—the crew rationed caffeine drinks like coffee and tea for the mission in order to ensure that we had enough for daily consumption. This did not cause discomfort for us, but for a long duration mission such dependence could be a liability.*

[HERA] *For myself personally, the continuous video and audio recording is what caused me the most discomfort. If I had frustration with someone in mission control, I would be mindful to not say anything on the radio to reflect that negative energy. However, knowing there is background monitoring I also did not feel like I had the same freedom to release that frustration. For example, after getting frustrated I like to do a very brief breathing exercise. With the ongoing video recording I felt doing so would show my frustration when I preferred to keep it to myself. I suppose this was part of the HERA experience, but in general I believe it is key to have areas for crew to go where they have the necessary privacy to recharge.*

[HERA] *The time enroute to completing our mission was psychologically engaging, while the time post-asteroid-rendezvous seemed much more monotonous ("lets get back home"). I have given talks on my HERA experience and one of my final slides addresses what I thought I'd miss and didn't; and what I didn't know I'd miss and did. In the latter category were flowers, plants, and seeing the sun in the sky.*

[IBMP/MARS 500] An unspoken conflict between West European and Russian test persons. Different engagement in tasks under experiment from test persons. [. . .] To avoid inter gender conflicts only males were selected. This was a missed opportunity to explore if or what gives conflict in mixed crews. Large differences in focus points for public relations departments in the agencies and the scientific crew [. . .]

[ICARUS] *Repetitive tasks, monotony of day, no access to outside world, designated times to shower.*

6.3.1.2 In-Situ HABs (Fig. 6.5)

Fig. 6.5 MDRS Q5 responses

[MDRS] *For me, the main issue was the way the bunks were designed. Each bunk room is private, but the beds are like bunk beds in that one room has a bed on the bottom and the room beside it has the bed located above the other bed. Due to the construction, wind outside, etc. The materials tend to creak and groan under stress which can be a little unnerving when you are sleeping with a 200 lb human only a couple feet above you. If one was claustrophobic, this would also be much less than ideal (although an ICE would be a bad idea anyway)! To fix this, I would keep each bed on the same height so that people aren't jammed over top of each other. Another issue was the stairwell; due to its steep incline and painful material selection. The stairway was un-climbable without shoes on because of the "gripped" metal that was chosen. Without shoes, the grip cut into the skin and was very painful. The steep grade also made it dangerous and led to multiple falls from crew mates during our mission. This usually occurred at night when a bathroom trip was being made to the toilet (which is located on a different level than the bunk rooms). I'd improve this situation by using proper materials for the stairs and changing their angle to be more in line with a standard stairway.*

[HiSEAS] *The walls of the habitat were white. I think some colors, at least in one room (maybe some nature-like decoration?) would have helped. The walls were not sound-proof: I could almost always hear where the others were, and what they were doing (except if I or they were outside on EVA, or in the storage container—the Sea Can). That made it heard to have some time off: where we were, and who we were with, was always reminded to us. The desks were shared and aligned against the outer wall, so we had our backs to the main room, which was somehow unpleasant.*

[MDRS] *We had a propane leak that was making two of our six-person-crew dizzy. I would suggest an alternative method of heating that doesn't pose as much of a fire hazard (perhaps radiative floor heating) and replacing wooden the wooden staircase to the first floor with a non-flammable staircase. The walls were quite thin in our hab so sounds would permeate heavily, such as flushing the toilet, or snoring, or people waking up at different times. I would suggest creating somewhat sound-proof rooms so crewmates could have a sense of isolation from the rest of the team to unwind and sleep to our needs, etc.*

[HiSEAS] *Granted we were living in full gravity and I enjoy camping, so aside from the chill during the winter [...] there wasn't much physical discomfort. [...] Social was tougher, because we had such a reduced pool of people to talk to and interact with. Relations with the rest of the crew had their down points, and we definitely had a couple people that were just a constant chore and shouldn't have been on the mission. [...] I also saw friends and family start to write less and less as time went on, which I figured would happen due to my own research on the subject but knowing you're gonna [sic] get kicked in the shin doesn't make it hurt less. Psychological is hard to tell honestly. If you're crazy would you know? I was also told by friends upon my exit that I seemed quite unchanged from the experience. [...] We used the living room/dining room for virtually all social interactions, only place that was large enough. we definitely would break into pairs and watch shows/ work in the airlock, biolab, or rooms. The hard part was establishing social norms for where public/private areas were to be, as the use of the place meant that at certain times of the day, public areas were private, and vice versa. Trying to fit all the complexity we could in a fairly simple space.*

[Tektite II] *The Tektite II Habitat consisted of 2 chambers of 2 floors each, connected via a crawl-through tunnel/passageway between the upper chambers. Interior volume was minimal but adequate for recovery/evening functions after diving for the daylight hours. The inhabitants of the 2 (of 10) missions of which I was involved were professional ocean scientists & facility engineers with professional approach to accomplishing their mission(s). Their issues were mostly physical with the habitat environment and its operational constraints challenging their needs for collecting & processing results.* [. . .] *The habitat required climbing stairs from the work level to the off-hours level and transferring from one chamber to the other through a small diameter tube, both of which was mildly irritating at times.* [. . .] *NASA used Tektite II as an analogue for Skylab relative to Habitability. [. . .] So, items & equipment were supplied for off-hours recreational time to promote habitability. A primary NASA objective for Tektite II was: "Generate research data on human behavioral dynamics and habitability assessment for small crews in confined living spaces." This was accomplished by recording each crewmember's activities every six minutes for the mission and regular questionnaires on emotional & social feelings at the moment (1). A key finding from the post-mission interviews, later verified with closed-study tests conducted at the University of Alabama in Huntsville (2), was that realistic work for the participants was critical in providing an acceptable level of "habitability." The two were directly related. Though books and discussions were significant for off-hours time, useful, productive work won out over all the games, books, audio tapes, movies, and special foods provided for recreation. At the end of the day on Tektite, the aquanauts, after having been out in the water for 6–8 hours watching their fish, etc. and consuming an evening meal, would often find themselves for entertainment looking out the windows and the cupola of the habitat. In fact, the cupola was so important for Tektite, the NASA people put one in as a Space Station requirement when the original NASA Space Station "skunk works" was first created in 1984 at the Johnson Space Center. One was eventually provided to ISS and still remains as a place the astronauts wish to spend free time, quiet time, as well as work time. Some crewmembers, notably the women crew, found the constant observation by the external sources as an irritation or distraction. Other comments alluded to the external observation as being a bit comforting. LESSON LEARNED: Provide internal volume & utilities to support anticipated & planned operational functions for the mission. LESSON LEARNED: Off-hours items are good to have, but incorporate a window, preferable cupola, in the habitat. NOTE: This may be a problem for deep space travel, as actual vision will be of a black surface with white dots. May consider using projected enlarged images of actual space view.*

[MDRS] *Poor sound insulation—being able to hear team members snore and bathroom sounds would disrupt sleep. Also we rotated cooking and cleaning duties to keep it fair, but should have assigned cooking duties to those who were better cooks. Asking someone who has stated that they are not good at cooking to prepare a meal does not have good results for anyone.*

[MDRS] *The single composting toilet not being designed for constant load; use two toilets for redundancy, and/or use a more appropriate design.*

[Military Training Base/MDRS] *The weather: In the case of the military, inherent lack of adequate climate control equipment was the most challenging aspect of the environment. No air conditioning in the heat or operating in extreme cold environments was a physical challenge as it was a psychological challenge. Some people who could operate in "normal" (room temperature) environments sometime found it difficult to do the same tasks when in a stressful (climate related) condition. Hygiene facilities: such as shower stalls were limited to none. Living facilities: Lack of privacy and "personal space." Sometimes people would be tired of each other with nowhere to run".*

[FMARS] *Difficulty of access to bring scientific equipment—lack of or poor internet connections—difficulties in communicating along the hierarchical line.*

[MDRS] *Lack of or poor realism of habitat w.r.t. some functionalities—abuse of time spent in chores of maintenance and repairs—nonexistent data storage and retrieval system for biology and geology samples—lack of stowage room for biology and geology samples—sand invasion in habitat during desert storms; cold at night as heating system failed several times*

[OeWF Analog Astronauts] *The water tank and sewage was at close proximity, so this had to be addressed during week 1 pre isolation. The medical area was not confined at the beginning and this needed addressing. There wasn't a locked refrigerator for the med drugs in a desert environment e.g., propofol, certain abx. We got this after a few days.*

[OeWF Analog Astronauts] *We had what we took or planned for. You can't just head to IKEA and buy that extra lamp you need for reading. The distinct lack of plants and green in the environment affected the team. It was better once the rocket salad was growing.*

[MDRS] *Temperature. It has heating and AC but the desert undergoes extreme changes in temperature that are hard to balance.*

[MDRS] *There is a limited water supply and diet, which are to be expected on long duration space missions. The diet is usually supplemented by vegetables from the Greenhab which we didn't have given we were the first crew for the season, but later crews would have. When new crew members arrived mid-mission, we had them bring some additional treats to satisfy people's cravings, but it was interesting considered what we needed and could get by on versus what we might want. Some of the service systems were noisy, which restricted their use at night. The presence of crew members who were skilled cooks made eating from a limited pantry much more enjoyable, and all the crew liked inventing new dishes.*

[BIOSPHERE 2] *The actual habitat was very comfortable and large, while I can imagine an uncomfortable habitat exacerbating stress on the crew, the physical habitat cannot be relied on to alleviate problems. I spent years on a ship that was not comfortable, similarly, the discomfort make stress worse due largely to sleep discomfort (cramped, cold moving bunks), but was not the source of crew relational issues.*

[MDRS] *The most inconvenient aspect for everyone, I believe, was the incinerator toilet that everyone took shifts cleaning. The next might be when one of the team got sick for 4–5 days in the middle of the mission. Tasks had to be reorganized real time and there was no leaving the facilities unless it became a life-threatening emergency—this was just a bad cold.*

[OeWF] *Lack of space/Battle for working space: having a structured allocation for workspace remedied that a bit; plus identifying the "clean nerd"—the person, who is most likely to pick up the cleaning tools and start cleaning. [. . .] We experienced heavy weather situations (sever dust storms etc.)—having to hope that the structures can withstand the unusual conditions is a stress factor. Imagine going camping and your tent gets destroyed—that is annoying and you might freeze a little. During a mission, it can mean the end of a project prepared for years by literally hundreds of people. You know that Mission Support is keeping their fingers crossed and provides you with excellent weather updates and some tech suggestions, but mostly, it's about "sitting it out".*

[Oceanic Vessels] *The lack of any single quiet place where you can reliably be alone/out of line of sight of people. Really the only location was your 'rack' (bed), but this wasn't even a possibility for a large portion of the crew who had to 'hot rack' due to space constraints. This meant that 2–3 people shared the same bed, and while one person was sleeping, the other was on duty, and they would switch out (3 rotations). Although not explicitly forbidden, it was also a taboo to be in your rack when it wasn't your 'oncoming' (immediately before duty) time. One way of improving this would be to rig up temporary 'wall' curtains in certain areas to create personal space. This isn't very feasible on a ship however.*

[Antarctica] *Comfort is an attitude. Troll Base is now upgraded, but the old base lacked training facilities which we improvised in a 20 square meter container. In addition, some discomfort/risk moving from one building to supply containers or aggregate house during dark winter and storms.*

[Antarctica] *No access to darkness. No privacy.*

[Oceanic Vessel] *I lived on an icebreaker around 120 m long and 25 m wide, with around a hundred other people. In that environment, your world quickly shrinks to the size of the ship. It becomes a microcosm with its own energy, rhythm and rules. The ship even works on its own time zones, shifting by an hour every day for a week as we cruised east—meaning we had to live on a 23-hour day. It felt like a novelty at first, but after a few days you begin to feel very peculiar.*

6.3.1.3 SPACE HABs (Fig. 6.6)

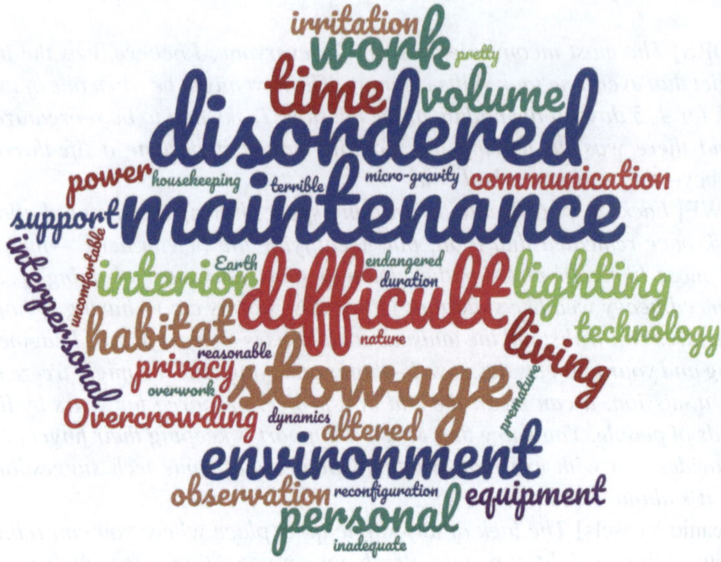

Fig. 6.6 All space habitats Q5 responses

[Skylab] *Due to power allocations and constraints of the time, the lighting was minimal (~5 ft cd ambient) in the Skylab habitat environment, making close work difficult. Supplemental (head-mounted) lighting was required for detailed work, especially maintenance work. Modern lighting technology produces greater illumination for the same power [. . .]. Additionally, the interior pressure of the 75% N2/25%* O_2 *environment was only 5 psia. Crew members often had to shout to communicate with someone 20 ft or greater from them. Initially requested but concepts not accepted was a maintenance workbench. Various maintenance workstations have appeared on Spacelab and ISS. A glovebox is imperative for a proper maintenance workstation. Artificial gravity of some degree would also be beneficial in controlling small items not planned for micro-gravity maintenance.*

[ISS] *Early issues have pretty much been resolved. However, initially, the personal issues dealt with need for laptop computer support. Initial units were not permitted onboard until they met flight worthiness constraints. Hence, the crew had difficulty communicating with the ground for family, technical, and maintenance discussions. Also, set up/teardown time for science activities was an irritation. The was part of the justification of flying the "Robonaut" robotic humanoid support device. It has not panned out but in time will be useful for mundane setup activities.*

[ISS] *In observing life on ISS, the most uncomfortable observation for me is the terrible tangle of cables, the untidy and messy nature of the interiors, and the extreme functionality with little regard for aesthetics. Living in a disordered and messy environment must have a psychological impact. Cables are a difficult challenge as there is so much legacy wiring that can't be altered, necessitating a very ad hoc / accumulative method of changing or altering equipment, but there has to be a better way!*

[Shuttle] *Too much volume common to all individuals compared to too little (or even no) individual privacy volume. Everyone has his/her own personal way of handling their own personal items. Before my first space flight—crew of 6 including 3 rookies—we were advised each to identity on-board the ship our own mini 'cubbyhole' where we could each store our own items used for daily activities. When you don't do that you tend to have your own items stored here and there in an non-organized fashion, which can easily irritate others.*

[ISS] *The ISS habitat living environment is characterized by very "full" and "packed" habitable volumes with lack of privacy (affecting interpersonal relationships and group dynamics). The view of the Earth from the Cupola is essential to make the crew feel "not too far away" from the people they love. An internal reconfiguration would be needed should the ISS infrastructure sustain not only crew members (fully focus on payload operations and housekeeping) but also future space tourists.*

[Skylab, MIR, ISS] *Poorly established stowage and inventory management systems; difficulty in finding needed items in a reasonable time. Overcrowding of stowed/stored items so that you cannot move or cannot work without moving things out of the way, and tracking where you've moved them to. On Skylab, the "jack-in-the-box" effect meant that as compartments were opened items floated out and it was difficult to get things back into the same volume they occupied when properly stowed. On Mir it was difficult moving around. You could often not see more than a couple feet in front of you; you could not take pictures of anything from more than a couple feet away. Finding things was often difficult and sometimes impossible. On ISS prior to the first long duration expeditions stowage had been dropped off on many earlier Shuttle flights. Tracking and finding where items had been placed was often difficult. Equipment not adequately tested prior to the mission. Exercise equipment like the running surface of the Treadmill on the ISS station collapsed in its first few days of use because of inadequate testing; endangered the mission and could have required a premature return.*

[MIR] *For me, too little to do. For my Russian crew mates, overwork.*

6.3.2 Experienced Positive Impacts of the Living Space

(Q6) What physical, social, or psychological aspect(s) of the habitat living environment listed above most positively impacted you or individuals or as a group?

6.3.2.1 Controlled/Laboratory Simulation HABs (Fig. 6.7)

Fig. 6.7 HERA Q6 Responses

[HERA] *The crew had good interpersonal relationships and no real conflict, so the physical proximity and lack of privacy promoted a sense of togetherness even when not socializing. Little teamwork was required for most mission tasks, but the few that required all crew to participate and communicate meaningfully helped to enhance the energy level and mood and gave the crew goals to work toward with objective feedback.*

[HERA] *The galley area was great. That's where we got to eat together, watch movies, have good conversations either personal stories or about the activities we were given to do during the simulation. Those talks increased teamwork and participation.*

[HERA] *Having a well-defined living/kitchen area definitely gave the habitat a homey touch, facilitating social interactions. We prepared the meals together, we all sat together at the table for all of the meals, and during time-off we watched movies together there, or simply (and silently) read or wrote, enjoying each other's company. This helped us to relax, which had a positive impact on performance. The sleeping quarters were extremely comfortable, and we all had excellent rest, which helped us to maintain an always positive attitude/mood, which had a direct impact on performance and wellbeing.*

[HERA] *It was very helpful to have the sound-isolated airlock in which to conduct some of the cognitive tests (the silent location was helpful for some of the more challenging tests). It was nice to have two separate levels. That way a couple crew members could be doing something that made noise (i.e. listening to music while exercising) while the other two could be on the other level without being too heavily impacted by it. Although our mission imposed chronic sleep restriction throughout the mission time, having private crew quarters to retire to each night was nice.*

[HERA] *Our HERA crew initiated "Formal Saturday" dinners. These were great for morale, even including the crewmate that was at other times unwilling to interact. I loved the velcro covering [of] many surfaces and items. It made it wonderfully easy to store pens, mugs, notepads, other small items that we wanted quick/easy access to. Weekly conversations with the psychologist, our family, and the medical doctor were all nice, as they gave us some variety and connection to the rest of the world. During downtime, we played games (scrabble, charades) and watched some television shows, which was nice bonding. Overall, I enjoyed not having to plan my own schedule or menu planning/food shopping, etc. I found it relieving to not have to make these kind of daily decisions and being fully directed by NASA MCC. (This was balanced by occasional frustration at not having the freedom to do what I wanted). […] We made a wall of artwork—hanging our own drawings. Helps make it a bit homier.*

[HERA] *I actually found the HERA hab to be fairly spacious; it was a comfortable volume to occupy for a month, with plenty of storage for just about everything. The sleeping quarters had privacy screens during our mission, and there was plenty of room. If I had some time and wanted a little quiet distance to write in my journal, I could usually find a spot somewhere. The temperature in the bunks was a touch cool, but tolerable with the right sheet or blanket. […] Lights in the second-floor galley/*

"living room" area had several different settings, including different color temperatures that helped somewhat with diurnal regulation, especially winding down after dinner. The bathroom module was big and well-ventilated, which really helped to dry out my sneakers and exercise clothes when I set them over the vent as I showered and minimized odors. The central module of the hab is a vertical cylinder with the elevator/ladder combo at the center as the long axis. It's hard to walk in a straight line for very far in the hab, but on the bright side if there's an obstruction blocking you from going in one direction, you can usually go a few extra steps around in the other direction and get where you need to be. The cameras and microphones all over the hab became a cultural touchstone for the crew, and an occasional source of amusement. We often cracked jokes about perceived or imagined responses by mission control upon observing our actions or remarks, complete with exaggerated personalities and voices [...] On another occasion, I had to relocate a camera to support an upcoming experiment and needed to stick around in the airlock for a few minutes to correct the field of view (i.e., "tilt the camera back up, now to the left a little, no, your other left," etc.). We had crafted a cape and crown out of leftover shiny food bags for a crewmate who had a birthday during the mission, so I turned the effort into a bit of a fashion show.

[HERA*] We deeply enjoyed our group time, either in leisure or during mission objectives. First, although the group setting in the morning could lead to dysfunction, for the rest of the day, it was a source of pleasure. Eating lunch and dinner together, while watching some movies was very fun. In addition, most of the mission relevant tasks were done as individuals or as a subset. There was one simulation that included all of us. It was not only rewarding individually, because it was just a fun simulation (I piloted a ship while others used virtual reality goggles to simulate space walking), it was rewarding as a team. We had to come together to do certain tasks the most efficient way possible. These situations were always a refresher and helped us get through the other times.*

[HERA] *This may sound contradictory, but it is important to have tasks to do during the day, as well as having down time to relax. Having individual tasks made time move more quickly, and having down time gave us the ability to relax. I looked forward to having meals at the end of the day with my crewmates.*

[HERA*] Movies, games and exercise! Work in HERA revolved around maintaining the "spacecraft," prepping for the big spacewalks when we would arrive at our simulated asteroid and building little robots. While ALL of those things were a lot of fun, my group really had a lot of fun unwinding in our off hours with games and movies.*

[HERA] *Privacy—The ability to get privacy in our crew quarters and in the waste management compartment was extremely helpful psychologically. It allowed time to escape from the other social divisions, to rest, reflect, and refresh. Privacy should be included in all future isolated long duration habitats in some form. Exercise—The ability to exercise regularly helped to keep everyone's moods lifted and energy up.*

[HERA] *Two things that come to mind are the ability to enjoy a meal together, and to have enough space to perform different tasks without interfering with each other. The shared meal space was certainly a positive for me and I'm confident it was for my team as well. This allowed us to enjoy fellowship while comfortably enjoying our meals. Thus, we were able to build our relationships and strengthen our bonds which in turn motivates us to support each other's productivity and* well-being. *Sometimes we would be performing unique tasks at the same time. Some required the ability to concentrate, others the ability to communicate to other team members. By being able to work in separate areas we could isolate any noise or conversation that had to take place to prevent interference with each other's work.*

[HERA] *One of the secrets is to identify the strengths of each individual and positively engage and reinforce those. Another is to not form sub-groups.*

[HERA] *Group assignments where we worked together to solve problem. Meals together.*

6.3.2.2 In-Situ HABs (Fig. 6.8)

Fig. 6.8 MDRS Q6 Responses

[MDRS] *For me, the shared common space was the most beneficial. This is where we would huddle together to go over the day, eat meals together, play cards or games at the end of the day, etc. The time for crew decompression and bonding was essential to the success of the mission and the continued building of relationships. The separate science dome was also very positive. It allowed a 2nd location for a crew member to go to get work done, have a bit of privacy and come back to the full group ready to participate fully. I did all my research in the science dome and I was able to listen to music and look out the window while I worked which was a great morale booster.*

[HiSEAS] *As we were only six and constantly around, we did become very familiar to each other. It was nice to always have someone to talk to about anything.*

[MDRS] *Cooking together was a lot of fun. It served as a place to come together regularly in a positive light. I enjoyed the close quarters of the kitchen as it forced us to interact and be social for a short amount of time every day. That was good for team morale. I would also say that being in the Greenhouse was a really positive experience. Attending to plants was great. A nice physical activity and you could see the growth from your efforts.*

[HiSEAS] *I think the EVA's that four of us were always doing definitely helped us coalesce. Finding new caves, sharing the sweaty suffering of the suits, going over cool pictures and such when we were back was always fun. [...] I would say that some group activities were a boon generally, like the games, but others were hit or miss. [...] Finding out what people are going to actually go for and end up appreciating would be great. No pressuring us to do stuff we hate. The habitat itself was fine for me, I liked the double height ceilings, and while the architecture could have been far better done, it didn't feel cramped in the living room if nothing else. Toilets that weren't such a nightmare to clean would have been appreciated too but what can you do.*

[Tektite II] *Coming in from the water the crew entered into a dry volume where water-interface work could be accomplished without having to move into the rest of the habitat volume. Quiet time could be found even in very close quarters in the cupola and semi-spherical bubble windows.*

[MDRS] *The large, domed-ceiling wardroom/common area/computer work area on the upper floor [was] very conducive to positive crew interaction, not claustrophobic at all, adaptable for multiple purposes (EVA planning, media interviews, food prep, eating, socializing, playing games, watching movies, individual computer work, ...).*

[FMARS] *The sense of wilderness in the Arctic and permanent potential danger—feeling of being first explorer in an uninhabited Arctic island—first users of the FMARS Habitat, so we spent since some time cleaning and finishing up the installation.*

[MDRS] *The use of the greenhouse for experiments.*

[OeWF Analogue Astronauts] *The central crew areas like the food tent [and] the exercise/gym area.*

[MDRS] *The use of energy and environment resources has impacted the crew dynamics in term of cohesion, decision making and frustration.*

[MDRS] *There was an effective demarcation between the work areas on the lower level and the living areas on the upper level. The upper level living room has a very cozy feeling, with well-placed windows allowing distant views as well as overlooking the immediate vicinity of the complex. The main table upstairs was used for meals, recreation, meetings, and work. The upstairs layout worked in favor of group cohesion. The provision of individual bedrooms (staterooms) and also outbuildings including the Science Dome, Engineering Building (RAM) and Greenhab provided ample opportunities for crew to be along if they needed to be. The bathroom area was well separated from the general living area, affording maximum privacy.*

[Biosphere 2] *The actual habitat was very comfortable and large. Good sleep conditions and privacy was important. Also, the ability to have private conversations with a subset of the crew or another individual. The ability to have private conversations with people "outside" or in mission control, especially a "champion" outside.*

[MDRS] *Group meals once everyone was somewhat comfortable with the cooking rotation, style, talent, etc. [. . .] This is an interesting component of small, less than a dozen, person teams because larger than that people may group or separate more readily. There is an ideal size in relation to schedule and space (i.e., volume) of eating locations and food preparation that keeps the group more cohesive—don't know what that optimal point is though. 2) the second was going on EVAs because that gave a relief valve to people, whether they needed it or not. A chance to go explore, to see new things, to isolate themselves a little or to share new adventures.*

[OeWF Analogue Astronauts] *Personal space—the ability to withdraw outside duty hours—even a small bed/bunk is enough if it acoustically decoupled from the rest of the crew. Everyone develops their own sweet spots, which are respected by the others. Joint space: the cantina/mass is always the social hotspot, so once people arrive there (even if it is in 2 m walking distance), I got more of a "walking into a restaurant-relaxed feeling", because I knew, that we would do social things and not focus on tech or science unless we wanted to. Outlooks: the larger the window, the better; during some missions, we also put a binocular for everyone there to "mentally step out even more". I am pretty convinced that virtual reality windows can't fully replace that feeling. Safety overview: Having a command console where the station telemetry, IT, power monitoring etc. is displayed in a central location. That provides the sense of safety, knowing that all tech infrastructure is nominally operating. The contrast between the magnificent desolation of the huge desert vs our tiny high-tech station we call home. Nothing beats that contrast, especially under a starry sky during a nighttime EVA. It's almost like an oasis of life, power, water, food, communication—THE definition of an outpost.*

[Oceanic Vessel] *Having a common area that was always open for use/never reserved—for the officers this was the wardroom, and it was where we had our regularly scheduled meals and could always go in our off-time if we felt like interacting more. The enlisted crew equivalent was 'crews mess.' It created a sense of camaraderie and routine which made the time fly by faster and improved cohesiveness.*

[Antarctica] *South Pole Station's main facility provides a variety of common spaces for the crew to congregate. The galley, gymnasium, multiple TV/rec rooms all give support frequent gatherings of people thusly inclined. Conversely, the station also provides individual dormitory rooms for the crew members, which allows them some small territory for them to have some physical/mental privacy from the rest of the crew. For some jobs, including my science tech one, our workspaces in the main station and outbuildings afforded a level of privacy as well. This ability to throttle the amount of personal interaction while living in a relatively confined location is a boon to psychological well-*being *during a long deployment. [...] Though it could be spun in a negative light, a big aspect of being at South Pole Station that positively benefits one/the group (in my opinion) is the deep geographical separation from the rest of the planet/human population. It is a rare experience, and the mental cachet of being somewhere so exotic and singular was a major motivating factor for me and many of my fellow crew. Like they say in real estate: It's location, location, location!*

[Oceanic Vessel] *The companion of a good collaborator and friend is vital to maintain yourself and your team running a project on isolated places. Of course, rewards like meals that you prefer, snacks or even moderate drinks like beers, sodas or your preferred liquor on the leisure moments.*

[Antarctica] *Separate bedrooms. Good kitchen facilities. Recreation area (living room).*

[Oceanic Vessel] *Group activities are essential for morale. In the tiny microcosm of the ship, small events can easily be blown out of proportion. The way to avoid that is to blow off steam and socialize. I also went to daily yoga sessions*

6.3.2.3 SPACE HABs (Fig. 6.9)

Fig. 6.9 All Space Habitat Q6 Responses

[Skylab] *Having a "table" where the entire group could congregate proved to be an excellent utility. Meals were shared, meetings were conducted. Free time was utilized looking out the Wardroom window in the side of the S-IVB Orbital Workshop.*

[ISS] *The Cupola mounted on Node 3 (Tranquility) has provided hours of relaxation and scientific investigation to many crewmembers since its arrival in 2010. The Cupola incorporates an electronic workstation and crew restraints as needed. Additionally, the interior racks were designed for relatively quick changeout as equipment failed or was updated in technology levels. This capability has permitted the ISS interior to be maintained to latest design technology over the long term. Human interface design requirements were established and coordinated with all participating disciplines so that crew interfaces were standard, even though crew members came from many countries. Consistency is absolutely required throughout the station regardless of pressurized module.*

[SHUTTLE, MIR] *Having a common place where the crew can gather and share a good time promotes on board "socialization"; the crew enjoys being together to celebrate a successful task, or to discuss a topic affecting the mission. The more friendly and warmly is the place, the more positive will be the atmosphere. In the shuttle we only have the middeck or the flight deck which are not ideal. In Mir, the 'table' in the main central module was a good place for this kind of gathering.*

[Skylab*] In both Skylab and Spacelab, the real time collaboration between crew and ground controllers and among/within the ground control teams was unforgettable. Someone should have been allowed to observe and analyze the group dynamics within those populations during those missions.*

[ISS] *The Cupola of the ISS most positively impacted the well-*being *of crewmembers, providing expansive views and the ability to track features on the ground and in space and providing mind-changing and relaxing off-duty activities for the crew. For twenty years many program managers tried to dump the Cupola as a cost-savings measure, but the availability of the Cupola has made stays much easier and more pleasurable.*

[MIR] *The fact that we trained together and had already formed a bond as crew mates.*

6.3.3 How Future Extraterrestrial Habitats Might Differ (Fig. 6.10)

Fig. 6.10 Q7 How will future extraterrestrial habitats differ from terrestrial habitats?

(Q7) How do you think future extraterrestrial habitats might differ from everyday terrestrial habitats? How will these differences change human behavior or mindset as we adapt to these new environmental conditions? How will they affect the individual and/or group?

Extraterrestrial habitats would likely be highly space limited and would have to balance the need for some elements of individual privacy while also containing prominent social areas to discourage self-isolating behaviors that wound tend to degrade performance and relationships over long times that would include repetitive or boring tasks. Spaces for activities such as reading, movie watching, etc. in individual and groups would need to provide some level of comfort especially in environments with long periods of rest or idleness.

We may not have all the tools needed. We will need to learn how to work with what we have available and maybe repurpose certain objects, work areas, tools.

Future habitats, at least at the beginning, would be inevitable reduced in space (compared to the everyday terrestrial spaces that we are used to); and depending on design, habitats would be noisy, due to the machineries and equipment (although technology will be capable of cancelling machinery/equipment noise). I imagine the need of having spaces for leisure activities, relaxation, meditation, and exercising! And spaces where virtual reality (VR) might assist us to "travel" to places we enjoy on Earth, and to savor terrestrial colors and sounds we are more familiar with (a blue sky, a green forest, the sound by a river or the sea, a gray rainy day). I also imagine VR rooms as essential to communicate with the love ones on Earth, and to do a variety of (virtual) activities, like going to the theater, or skydiving. We would have to adapt to this virtual world, which we are already doing on Earth. I also imagine habitats where artificial intelligence (AI) will assist us in our daily tasks: from performing specific activities, to (already suggested) emotionally intelligent AI that could support the habitants in different capacities. We would also have to adapt to this co-existence, which, once again, we are already doing in some way.

Such habitats are likely going to have to be smaller than typical terrestrial habitats, and as a result of that, they will need to be designed with efficiency of space in mind. These tight spaces are going to require that crew composition is very carefully determined in advance, to hopefully decrease friction points among crew mates. The smaller living spaces will demand that crews learn to function as a team through good and difficult times, with mission success being the primary motivating factor.

In regard to crew mentality and habitat design, a 1-month mission versus a 3-year mission will be remarkably dissimilar, and so the answer really depends on the planned duration. One big difference is that the materials of a habitat have a very different feel compared to our regular homes on earth—no wood or brick. I can't honestly say if it is important to provide some decorations/designs that bring a remembrance of Earth, or if travelers will be happy to embrace a metal-based home. Given the psychological effects of being away from Earth/family, I think making habitats as physically comfortable as possible is important. This means ceilings that are tall enough without ducking (I'm tall and had to duck through doorways in HERA). Quality food is important. I think bringing some culture/art/dancing/movies/music is always a strength. Surround sound systems will be great. I'm excited about the potential to use different lighting systems (colors and intensities) to enhance the habitat environments. Our HERA mission was testing a lighting protocol that would help us retain a pseudo-circadian rhythm. It seemed to work well. In general, providing a habitat where crew members can be as self-sufficient as possible is desired. It is a bit of a bummer to take all the orders from MCC, and it's nice to feel self-reliant. It is especially important when the communication delays are long. I think a feeling of Earth-independence will undoubtedly arise in extraterrestrial habitats. I suspect there will be a lot of robotics involved in future extraterrestrial habitats.

I think habitable volumes will be smaller in extraterrestrial habitats, both for construction and logistical purposes. Human interactions will consequently change to be almost bimodal in their distribution of proximity: everything will either be pretty close together in person or remotely via telecommunications methods, with not much else in-between. That said, I think it would be smart to incorporate at least one larger volume capable of facilitating meetings of the entirety or large fractions of the total occupying group, to promote and facilitate communications and camaraderie; it's a necessary balance to the privacy a smaller compartment can bring, a place for privacy without isolation, where the kind of incidental interactions that strengthen group cohesion can take place.

You give up a lot of control by being in a terrestrial habitat. You put trust in the people who designed the habitat, the scenario, and the equipment you use. In an extraterrestrial habitat, you are further away from home, in a harsher environment, and in a smaller space, and you are putting your LIFE into other people's hands.

One thing that we tend to take for granted on Earth is nature. In the future, when teams are working in long duration habitats on the Moon, Mars and beyond, I have a feeling that the need for some kind of connection to home will be necessary. An Earth-simulator, if you will. While food, recreation, exercise and movies will be something that our future crews will need to unwind after a day at work, I think that some kind of sensory stimulation will be necessary as well...at least for the first generations. Anyone born and raised on Earth will miss the sound of thunderstorms, rain, trees, birds, etc. I believe that some kind of Earth-like visual or sound immersion will be wanted during a long expedition to an extraterrestrial habitat. Think of the holodeck from Star Trek! Even during a 30 day isolation study, we tend to miss the feeling of the sun on our face and the smell of fresh air. While some of these effects can be replicated by visual and audio immersion, they will only satiate the senses for so long. As we start moving out into space for longer and longer durations, we may have generations of people born having never seen the Earth. These necessities wouldn't likely be as important to these folks, but they may be a good way to prepare a person for their first visit to Earth...a way to let them hear some of what the planet will sound like when they open the hatch for the first time. I believe that this will help remind the crew of home and provide a necessary mental push to get through any adversity that they may experience. People have bad days when they are deployed, and a reminder of home, of family, can help turn their day around.

Future extraterrestrial habitats will continue to be smaller and more spartan than their terrestrial counterparts, especially those being used for exploration beyond Earth's orbit. This will make it harder for crews to enjoy the benefits of privacy, sleep, food, and physical exercise. Efforts should be made to provide low-mass, low-volume mitigations such as crew training and in-flight support for group dynamics and individual psychological health. Group activities such as games, movies, meals, etc. should also be planned for the mission. The ability to communicate with those at home should also be maximized (though not forced) even if real-time communication is not possible.

There will be a heavier focus on function over form. The space will have to utilize a small space effectively while maintaining an acceptable level of comfort. This will result is much less space, less luxurious accommodations, etc. It will create an environment that the home and workplace are the same. There will be less opportunity to have private space or time away from the same people. I have learned that my interactions and personality at work is different than at home in the sense that at work I need to be more direct and polished. I believe these elements will combine to require an additional element of mindfulness in order to maintain a healthy mindset. Even during the most stressful times at work one can just 'push through' the day knowing they can relax at home away from the work environment and/or coworkers. This will not be possible in future extraterrestrial habitats, thus requiring the additional need for mindfulness.

Much more mission focused environments, but that can be beautiful in its own way to the right individuals. Take the Apollo command module, for example. Spaceflight mission accomplishment will, to a great extent, drive future astronauts to accept the utilitarian environment that distance, cost, and mission dictate, but for long stretches so remote, and so far from home another dimension of habitat design will be necessary to maintain positive mindset.

We had access to unlimited resources such as running water and food supply. We also each had our own bathroom and bedroom. Regular access to personal space was paramount to maintaining social relationships in a confined space.

They will require their inhabitants to be much more intimately aware of how they are constructed for the purposes of maintenance and repair. In these cases, the users will have a level of intimacy and knowledge of the habitat that likely surpasses even the engineers/architects who designed it. As such, they will likely uncover quirks and design flaws/features that were not foreseen prior or intended. In order to flesh out the potential negative aspects ahead of time, the forecasted users should be closely involved in the design cycle in order to provide feedback and mitigate frustrations down the line. If this is able to be done, habitats will implicitly be much more "personalized" to the users and reflect their specific desires and personalities.

Future habitats will be more compressed than our traditional terrestrial habitats and this will have an impact on both individual and group dynamics. It will be more difficult to have private/personal space which can lead to increased stress and tensions among the group. People will need to be comfortable with less personal time and less personal space as we transition to this type of living arrangement.

They will be shaped by constraints on volume and weight and be optimized to an extreme.

Daily lives are completely changed when living in isolation. You can't simply walk outside, or have natural sunlight, or feel the wind in your hair. Being on another planet will really challenge what it means to be human. Living in small spaces closely with other people comes with challenges. Keeping busy is key, but also creating spaces for socializing, exercise, and being encouraged to visit a greenhouse frequently would all be things that I would assume a good hab would include.

I think that their response to available materials and gravity will dramatically influence the design. Printing from regolith is going to eventually lead us to a bunch of Gaudi-inspired lunar architecture, same for Mars. Depending on our ability to make them to a certain scale, the greatest influencer will be how much humanity we can pack into these things. The longer the duration the more we need what will seem to be superfluous stuff to keep people healthy and happy. I think people born in these habitats (eventually) will be just fine, but that people who need to adapt to live in them will have their individual troubles. I'd be curious to see the solutions that come from the extra unlivability of the environment. Less windows, designs that prioritize environmental security, issues with radiation….there are so many forms and responses to extraterrestrial habitats that the eventual design solutions could be almost anything.

The habitat will have to be lightweight & pre-integrated with utilities, communication interfaces, some level of flexibility in habitat interior appearance (simulating appearances of seasons?) Much key equipment and capabilities will have to be available on-site prior to crew arrival and require only minimal set-up after arrival. Natural exterior environmental conditions must be accounted for prior to mission initiation. EX: Lunar dust was a terrible issue on the Apollo missions, even for the very short missions. Such dust must be eliminated in the habitat interior, left outside. Water in and underwater habitat must be kept out of the living volume of the undersea habitat due to its impact on crew & equipment health. Relative to crew mindset, humans are adaptable to adverse or different environments. Behavior will follow mindset. With proper stimulation (e.g., realistic, useful work) can counter many gripes of habitat conditions.

Need to incorporate relaxation/decompression aspects for people, i.e., not just continuously work.

Extraterrestrial habitats will require the human inhabitants to be more resourceful than they may behave on Earth. In an environment with limited resources, human behavior will have to shift to be more economical and strategic with every decision and action.

Need to balance total mass and volume, durability, repairability, redundancy, habitability, cost, with safety-of-life as a priority, will make it challenging to also make them "livable". Especially since it's hard to predict human behavior, especially group behavior, under long-term isolation and risk, and so habitats should be designed for adaptability as well, so that the crew themselves can experiment and figure out what works best for them.

In extraterrestrial habitats, no room for error, especially on life support systems, would be an ever-present worry and could lead to zero-error tolerance, which could be detrimental and unbearable in the long run. A redundant system would be necessary. I think that the "permanency" of such operations would have a big effect on behavior. Are crews there on "a one-way mission" or is there a "return to earth" mindset. The latter would be the healthiest conditions (whether that is a couple of years, it is a different story).

*Future extraterrestrial habitats will need to respond to differences in the external environment including gravity, natural light, day/night & seasonal cycles. They will need to protect from hazards including dust and different types of radiation. They will necessarily be spatially constrained and highly adaptable. They need to be made as comfortable and attractive as possible to avoid imposing unnecessary stresses on crews. Provision of private spaces for all crew members will be critical for long term well-*being*. The ability to see outside the habitat and have plants somewhere in the living environment (even just to grow food) area also positive for morale. The living spaces should discourage crews from eating alone and becoming too isolated. Cleanliness and hygiene as important. In space, you can't go elsewhere if the habitat is causing discomfort, you so the design needs to avoid major problems from the beginning. That said, human culture is very adaptable, and small cohesive groups have great capacity to embrace and work within constraints.*

The habitat is like a super compact city, think of the functions we use a city for and replicate them. Facilitate many forms of communication with Earth. Facilitate private meetings just within the crew.

All space habitats will need to be highly designed and optimized for redundant, reparable, and efficient operations. But there are two kinds of space habitats, orbital and planetary surface. If designed and operated correctly, people will be able to conduct EVA activities much more rapidly and regularly on any planetary surface. This will help people adapt to the environment easier. Then again, individual history and personality do play a huge part in how any individual will respond. Those who grew up in more outgoing conditions, self-reliant situations, will likely fare better. For crew sizes smaller than about 20, this will require full discipline cross training and scheduling of facilities and janitorial and cooking activities. No calling a plumber when it breaks. Initially this might create a more cohesive group, smaller scale, because everyone should equally share in the goals and operations of the habitat—for a safe and successful outcome. This will likely change with much larger team sizes. Also, for crew quarters, there are differences between shared and individual that should be explored.

The reliance on more high-tech with more tendering requirements is challenging. If we have a structural damage [on Earth], no problem; if it happens on Mars, you're dead. So, the feeling of dependence on tech and structural design is a matter of life and death. So, if crewmembers get sloppy (e.g., with dust mitigation protocols), everyone is affected and at risk. I believe that will lead to an institutionalized slight "paranoia" of watching each other's compliance, even more like e.g., on a submarine or a sailing vessel. Some of the risks will be invisible, like radiation or long-term health effects by low gravity etc.—making sure, that this is well understood is more than a courtesy out of psycho-medical curiosity, but a must to avoid hypochondria and overguarding of each other. More Back-ups, back-ups, back-ups, contingencies, spare parts, redundancies. Did I mention more back-ups already?

Compared to anywhere on Earth's surface (including South Pole), extraterrestrial habitats may have to be designed intentionally with respect to whether they will attempt to provide any Earth-like aspects. Should a living space attempt to give the crew a reminder of the home planet or should it just embrace the alien setting and leave Earth to be a memory to look forward to upon return (if any)? On one hand ersatz reminders of blue skies and wind in your face could help reduce stress in some crew, but in others it might be a distraction and stoke unrequited longings that prove to be a distraction. Even if only a small portion of a (likely small) crew adapt poorly to whatever spectrum of Earth-like/other-worldly design aspects of their living space, this could greatly reduce team effectiveness and morale if some are having adaptation issues. Depending on the reliability of the habitats' systems, there might be much more of a tendency to think negatively about the habitat if there are frequent issues that pose risk to crew health and safety. South Pole Station is not an everyday terrestrial habitat, and the relatively infrequent (but reoccurring) issues that have been experienced with some of the systems there (e.g., the power plant) that have real and significant effects on crew peace of mind and workload certainly can color one's perception of the living conditions. If your life support system is questionable, it necessarily leads to questioning your chances for survival. Extraterrestrial habitats obviously will need to provide reliable systems to best ensure mission success, but this also will drive crew morale/performance.

I believe habitats will be flexible (especially interior) based on not only mission but also group preference/needs. Given timeline e.g., mission/project or longer duration/life, [the] habitat could be adjusted to needs in more effective and sustainable than today. E.g., a family needs a different habitat then a mission team or a couple. I truly believe we need/will adapt as humans. Especially regarding what's for someone is called "comfort" but for others are "needs". Basics needs (safety, food +++) will be there but crews need to learn more about how we are functioning as humans especially when it comes to consciousness and how it works and affect us.

Gravity. Not just staircase design, but the bigger ethical issues.

These spaces will have to be the inhabitants' whole worlds—their inside spaces, outside spaces, places to socialize, places to be apart from others. As humans we are used to moving through different spaces—our bedrooms, kitchens, houses, streets, public transport, offices, countryside, wilderness. The spaces will have to be sufficiently diverse that it can provide some of that necessary variability. A simplistic, monotonous space would lead to a feeling of being confined and trapped. Any design features that can prevent or alleviate that feeling will be essential.

Future extraterrestrial habitats in space will likely mix some elements where an artificial gravity is re-created and elements providing a weightlessness environment. Each of these types of elements will have to benefit to the maximum extent of the peculiarities offered by these two different gravity environments. Their interior design should be such that residents should enjoy transiting from one to the other.

One should not offer a better comfort than the other but offer a diversity each with its own attractivity (a bit like when you like to travel and see new places and after some time you like to be back in your own home where you recognize your own identity—who you are). Future extraterrestrial habitats on the surface of other celestial bodies (Moon and Mars in the coming decades) will impose living in an artificial atmosphere. So, there will be a permanent airtight structure maintaining this structure. For the same reason as above this structure should not be the habitat structure itself but should rather house (and protect) a set of different and separated habitats giving their residents the sense that they can go out at times. For both kinds of habitats (in the middle of space or on a surface), the law of physics will privilege spherical and cylindrical shapes. Rather than mimicking our flat habitats surfaces on earth (floor, ceiling, walls), individuals will have to accept adjusting to non-flat shapes. The reduced gravity will allow them to handle big and/or massive objects impossible to move under Earth gravity. This 'augmented' force capacity should be taken into account in the design of Moon or Mars habitats as it will offer plenty new possibilities for daily domestic tasks as well as for work tasks. A group might be affected if some individuals prefer to have their habitat mimic their former Earth habitat.

The initial habitat of the infrastructures to sustain Moon and Mars exploration will be characterized by a very limited volume and by the need to provide radiation protection in a very harsh environment. Basic needs to be provided include comfortable environment (colors, shapes, smells); internal layout reconfiguration; possibility to see outside; communications with families/friends; greenhouse to complement food; cooking option; personal crew quarter; social area for relax/exercise.

Future extraterrestrial habitat designs, and operational guidelines must acknowledge the need for the evolution of new norms and provide accommodation for them. I believe there will be strong pressure among residents to adapt and evolve hierarchical and social norms and policies in response to the extraterrestrial paradigm; so the 'system' must allow the inhabitants to tailor their norms and policies to achieve a comfort level and a perception of 'self-determination'. The 'we' (extras) vs. 'them' (terrans) distinction will eventually exist. it will be either a constructive distinction or a destructive one depending on how well this dynamic is accommodated.

Confinement! (because of generally inhospitable outside environments) I suspect that the problems will be more psychological than (directly) physical. Related: the same this on a broader scale: no escape! Limited (real-world) social networks, including limited ability to find time/space away from others. Corollary: limited sex life.

Different gravity levels will affect many aspects of extraterrestrial habitats. Without convection (or reduced convection) moving breathing air and CO_2 around becomes intentional, not a product of the environment. This also influences air cooled equipment. Routine and vigorous exercise will be required to counteract bone demineralization and muscle atrophy. Beyond the Earth, electrical power and life support functions are absolutely essential. Therefore, these systems must be *designed for reliability, repair-ability, and maintainability. That translates into logistics, storage, and resident skills to assure continuous operation.*

In terrestrial conditions we can tolerate limitations and dissatisfaction of our living conditions for some time as there is a possibility to move out or change the environment. That will be different on other planets. Seems personalization of a living space will be important for inhabitants and groups will be more cohesive to survive and be productive.

6.3.4 The Three most Desirable Characteristics of Future Living Spaces

(Q8) What would be the three most desirable characteristics of living spaces located in extraterrestrial environments? What part of the adaptation process would be addressed? And how could these be integrated/incorporated into an extraterrestrial habitat? (See Table 6.1)

Table 6.1 Question 8: Summary of three most desirable characteristics of living spaces located in extraterrestrial environments

Description	cnt	Architecture	Psycho-social
Privacy and personal spaces	26	Adjustability	Group dynamics
Flexible common socializing spaces	12	Adjustability	Monotony
Flexible lighting that can also mimic sunlight and circadian cycles	6	Adjustability	Monotony
Reliable systems	6	Function	Safety
Interesting views both internally and externally	5	Adaptability	Monotony
Configurable multipurpose spaces both social and work	5	Adjustability	Monotony
Decorative elements including color	4	Adjustability	Monotony
Good kitchen and shared space for meals	4	Adaptability	Group dynamics
Good exercise areas	4	Adaptability	Monotony & group dynamics
Effective communication with earth	3	Function	Safety
Dedicated and varied functional spaces	3	Adaptability	Monotony
Green space—real or Sim	3	Adaptability	Monotony
Efficient and flexible pathing	2	Function	Performance
Easy housekeeping	2	Function	Performance
Ability to create a space that reminds of home	1	Adaptability	Comfort
Effective workstations	1	Function	Performance
Effective storage systems	1	Adaptability	Performance

Some soft accents/surfaces particularly in berth areas seemed to help increase the sense of comfort and noise deadening. Design should reflect some attention to maintaining functional and efficient workstations, but individual quarters with softer but easy to clean elements would help separate work from personal spaces since those areas would generally overlap more in confined spaces that most people are accustomed to at home.

1. Good and effective environmental control system; 2. To be able to communicate back to Earth with less communication delay; 3. May be using virtual reality or having someway (maybe through playing something on the windows) creating an illusion of day and night and some nice scenery.

Multi-purpose—Small doesn't always mean better. Able to change colors around the habitat (this could be through paintings, colors, scenarios, etc.).

1. Spaces for socializing (which will address a basic need of human beings); 2. Spaces to cultivate the mind and the body (where to exercise, mediate, etc.), to maintain mental and physical health; 3. Spaces that assure a restful sleep to recharge. I think that spaces for socializing could be easily incorporated as part of common areas to cook and dine, and, in more spacious habitats, maybe "open" spaces, like gardens, can be built. I can imagine relaxation/meditation rooms as small/comfy "capsules" VR-assisted. Exercising facilities might require a more specific design, and if VR-assisted small spaces could be (virtually) transformed in bigger spaces, to perform a variety of sports (a static bike activity could be transformed into a mountain biking adventure, or a pleasant ride in the city, like some game software already does). As for a restful sleep, cozy capsules could be a solution. Once again, I see VR as a key tool, maybe providing the illusion of being in a bigger room, with the preferred setup of the habitant.

1. Living spaces should be designed for efficiency of movement for multiple crew mates. As an example, the circular design of HERA seemed to be an excellent design for such a small structure, as it allowed crew members to access locations from two different angles (i.e., if a crew mate was in the way, we could always travel the other direction around the circle to get to our destination); 2. In these tight quarters, having different isolated regions will be helpful (i.e., different levels of the structure, and private crew quarters). Just a little bit of physical separation with some sort of physical barrier between areas can really be beneficial; 3. Each space should have a designated function (i.e., an area that is primarily devoted to work-related functions, another that is devoted to living (cooking, eating, relaxing), and another that is a private space for the individual crew members.

1. Options for privacy—In HERA, our privacy was limited to the bathroom, our sleeping pods, and occasionally the airlock. It was about right, and I don't think it would be healthy for crewmembers to hide away for the duration of a mission. But some option is important, I think just to offer some personal time for reflection and decompression. Also, and to be frank, I've never heard anyone seriously discuss the topic of sexual frustration in space. Is masturbation just not discussed when designing these habitats, or are people just expected to ignore this topic while in space? 2. Materials/decorations/lights—Together, these can really impact the mood of the habitat. I recommend asking the crewmembers what their preferences are.

3. Convenience/ease of action. While increased space is desired, convenience is paramount. Systems should be designed to reduce frustration in finding tools, activities, etc. So, I guess storage systems should be a focus.

Having a good view is important. Lots of folks like the idea of being able to look out at the stars and landscapes outside. Sure, that's appealing, and I'd like to see such things incorporated into habitat designs. For me, though, I think it's equally important to have interior views, almost like the interior rooms with balconies looking over a grand atrium in a large hotel. You can see greenery, you can see people interacting, and you can choose to be sociable (curtains open) or private (curtains closed). Some folks may like an exterior view, while others may want a view of a greenhouse or arboretum interior; size and layout of the facility come into play. In line with this thought, having a personal, private volume is also important. I don't think "hot bunk" style habitation would be tolerable over a long term. While it may be too extravagant to consider even a small private suite and private bathroom for the individual, I think something shared with a spouse or a roommate or two is acceptable to most folks. I like the idea of reconfigurable spaces, but only to a point. I'm always wary of the workbench design with all sorts of fold-out tables and sliding shelves, because if I have to pause in the middle of a large project, I can't easily stow everything back away without significant time and tedium. If an area is to be multipurpose, let it be as general as possible, like a school cafeteria. It can be a cafeteria, but also a dance floor, or town hall, or triage center. Give significant thought to which activities can be multipurpose spaces, and which should have dedicated facilities.

As I mentioned before, I would need to feel like I was in control of something. I believe it would help if I had my own little area that I could customize. Whether it be pictures on the wall, or trinkets, or even able to move things around a bit, my need to control things could be soothed by having my own customizable space. It's not exactly clear what that space might look like, but maybe if it had modular pieces, I could make it more comfortable for me. Those pieces could be from the ability to adjust how I sleep to how I hang stuff or store stuff. And I'm everyone would want their space to be a bit different. I would also want a community area, one in which bonding, and camaraderie could be built. I can't say in specifics what that might entail because every group would be different. Maybe that could be modular as well. In this instance, modularity could be from facing each other to all surround a focal point (like a screen).

Common/recreational space; individual space; kitchen. [Would address] team building/problem-solving skills; ability to decompress; camaraderie. [how integrated/incorporated] Create the environment necessary to induce the behavior you want. So use expandable rooms or build the hard infrastructure into the spacecraft.

When I think of living spaces, I'm thinking of crew quarters, or something like that. They should definitely offer some privacy. Sometimes you just want to go off by yourself and think or read. While it's unlikely that someone with any kind of group anxiety would be on a long duration mission in the first place, everyone still needs their privacy from time to time. Initially, crewmembers spending some time alone in their CQ's could listen to personal messages from home, call home, or even just read and relax. This would alleviate some of the homesickness that can occur at the start of a long trip and help ease the crew into the right head space for their mission. I think it would be done like it is on the ISS. Have an enclosed room available with some soft padding on the inside that insulates from some of the noise. Personal computers, media devices of some sort, and a space available for pictures on the bulkhead. For habitats with some gravity vector, a minimum would be something similar to what we have on Navy vessels. A rack bed large enough to sleep in, store some personal items, and the ability to slide a door closed could be provided.

Private crew quarters, external communication, exercise.

1. Shared community space for meals, rest & relaxation, etc. 2. Access to private spaces that aren't necessarily all the same. Some may prefer lights in a different configuration, different sized spaces, etc. For example, I am also a firefighter and have found during night shifts at the station some prefer to sleep on the couch in the community room, some prefer sleeping far away from others, some do or do not like windows, etc. 3. The capability to be modified. Different crews or even experiments may have different preferences on the utilization of a space. For example, one crew may prefer a larger community space and smaller private spaces, or vice versa. These characteristics would address the process of being able to allow both extroverts & introverts to thrive. It would allow for flexibility as the composition of crews change, or as the dynamics of one crew changes. These could be incorporated into a habitat by allowing for a modular internal habitat design, like a cubicle space in an office. The barriers creating cubicles (which could be tall enough to be walls) can be easily configured to suit the needs of whoever is occupying the space. Even in HERA, which had very little room for customization, could be potentially modified to some extent. For example, some crew members may prefer to remove the padding between their bed and the shared space to create a more open environment (like a loft).

From my experience, integrated into long-duration habitat, a place to escape to with a piece of home will be important. The psychological benefit may outweigh the initial cost. My thought would be something like one room with a wood paneled library, a simulated fireplace, and plenty of plants to care for and personal memorabilia. Something the group could design and agree upon prior to mission execution.

1. View to outside world; 2. Private personal spaces; 3. Group space for meals.

Dedicated private spaces for each person that block off line of sight, even if they have only a soft/non-audible barrier. A dedicated communal area that serves no other operational purpose other than socializing and at most meals. Dedicated exercise space with enough unobstructed area to allow full range of motion if possible.

(1) A personal area to sleep/read/be away from the crew; (2) A shared common space for team interaction and team building; (3) Windows to see outside of the habitat. During my experiences, it's the human dynamics that are the most difficult part of the mission, not the confinement. By designing personal space and shared space, it should give the opportunity for people be alone when desired while also bringing the crew together to talk, learn, discuss and achieve success in their

mission. This could be achieved by designing personal crew quarters that are private, but that are then attached to the shared space for the crew.

They should help counter the sensory deprivation that threatens crews in such missions, for instance with colorful (and possibly changing) decoration. They should foster casual time spend together, for instance with a pleasant and convivial equivalent of a living room. They should allow for some private time: a personal space, even small, where one cannot hear or see their crewmembers and work environment, so they can take a break from it all when they need it.

(1) Sound isolation for individual teammates—to really give deep sleep, and noise-free down time, un-interrupted personal time to reset from a day's work; (2) Comfortable social areas—like a space to interact with your crewmates in a non-work matter, say with comfortable couches or chairs; (3) I am torn between saying a greenhouse (but I assume that isn't a characteristic of a living space and more of a requirement of a hab) and an exercise room. (1) soundproofing—foam—airspace in between walls to dampen noise—or noise canceling technology; (2) Not sure how NASA could transport a couch to another planet but I would assume inflatable couches would be a reasonable thing to provide comfort and relaxation to crewmates; (3) Since exercise equipment is obviously heavy, providing resistance training equipment (like stretchy bands) would be a simple and light way to encourage exercise.

I would want lighting that feels as close to real sunlight as possible, and changes hue and intensity throughout the day for my circadian rhythms. Lighting is a big deal. I'd love to have real/simulated nature in the space. Garden walls with screens behind them that simulate a great distance. Hidden speakers with nature sounds. Running water? A single live rabbit that will cuddle with you randomly when you sit on the one patch of grass for tea? Even if it's just an awesome small room with VR and good speakers to enjoy simulations of nature from Earth. As cool as they look, I don't wanna [sic] live in a sterile sci-fi place. Habitats should have faux wood, nice stone, personal art and fun knick-knacks. Feel closer to a home, hostel, boutique hotel as compared to an institutional complex. If I'm living for three years on Mars, I wanna [sic] kick it in a cool bed and breakfast that feels as though it were carved into the landscape. Not a hospital or motel. All of these could be incorporated in a variety of ways. Decent 3D printing of local materials, Hue lights or Coelux skylights, good speakers, environmental systems, even just spending the energy to bring along some real materials so one room of the habitat feels like home, whatever makes it work.

(1) Proper and controllable lighting; (2) flexible paneling of compartments; (3) Access to privacy area (even if it is only a temporary and flexible space). Simulation of living spaces can offer potential issues which must be addressed once on site. However, the actual extraterrestrial living space will offer conditions not available in simulation—odors, textures; delta gravity levels, etc. So, flexibility of living spaces will be a boon, as actual time in the environment and team members will offer corrections from Earth-based sims., including illumination, acoustical modifications, dampness control, changes in the environmental scenery possibly offering scenes past experiences on Earth. Proper flexibility of habitat spaces,

textures, acoustical control will offer some level of adaption to what one would have brought from Earth experiences. Multi-use materials, equipment, etc. can help in providing such flexibility. Some amount of personal separation is critical for crew members, even if for short periods of time. The habitat must support the work to be accomplished; any failure to do so properly will result in major issues perceived with the habitat.

At least a small bit of personal space for each individual, a place to store photographs of loved ones etc. Space needs to be functional and designed with the human user in mind, adapted to the specific environment.

1. Having a small private space to allow for some solitude; 2. Ensuring the living space has a source of light and vitamin D—windows for natural light or artificial depending on the location in the universe; 3. A space or equipment for exercise—to keep in shape and create a space that allows the inhabitant to relieve stress.

Spaciousness, homey feel, adapted to the patterns of everyday life (i.e., efficient layout that minimizes wasted space and wasted travel, but also allows for privacy and room to explore, and fosters both serendipitous encounters and opportunities to withdraw for some private time).

Individual sleeping quarters where personnel can isolate themselves if/when needed (or to quarantine them). It'll give personnel time (and space) to gradually adapt to the new challenging environment. Appropriate lighting for circadian rhythms (adequate sleep); Adequate climate control (comfort); View to the outside (prevent claustrophobia).

Functional but with minimum of comfort allowing crewmembers to fully rest; constant power availability for all habitat functions; connectivity, internet to outside world and Wi-Fi for intranet and internal connections.

1. plants/green spaces are important; 2. Maybe a VR with scenes of home; 3. Maybe a mindfulness space with music or water.

Display technology, opportunity to reorganize the indoor facilities, keeping temperature and ventilation as constant as possible.

Sensory comfort—the small things really matter if you are living in the same space for months or years at a time. This includes heating/cooling, acoustic, olfactory, visual comfort (attractiveness). Make the hab easy to keep clean. Control of privacy—living in close confinement with others can be stressful and crew must be provided with private spaces they can retreat to. Reduced feeling of confinement—could be done by maximizing open spaces where possible, including outlook to the external environment, even if through only small windows or reflected using mirrors.

Enable privacy; Facilitate group meetings/interactions; Facilitate external support from the "ground"/mission control for the group and individuals.

(1) The ability to have a portion that is personalizable and private (but then again this might be culturally dependent). This reduces stress, increases comfort and a sense of belonging. (2) A commons space, for smaller groups, that can support eating and leisure actives, separate from work areas—optimization of such a space will require details on group size and gravity environment and other things. (3) If things are for very long duration, possibly the ability to modify the visual aesthetics

over time or on an individual basis. (4) Make traffic flow safe, easy and useful. (5) Easily cleaned. Every time the environment is designed to by-pass human thought processes (i.e., user friendly), it reduces stress and increases comfort in the environment (often without the inter-actor's awareness).

Ability to switch between private spaces and social common spaces: acoustical/vibrational/visual decoupling between those. The more space, the better; adjustable multifunctional spots (foldable trays/tables, curtains, inflatable external structures to increase living space, digging underground for storage etc.; use the vertical dimension; in most stations I've lived in, they are "quasi—2d" (even the MDRS); ladders, wall mounts, use the space below roofs, everything to increase living space. Centralized monitoring and control station—so we know about the conditions of the station at any moment. Data streams to sooth the mind that it is alright, and that vibrational noise is nominal because of XYZ.

1. Provide a feeling of security that they are a dependable place of refuge that has the tools/features needed to support safe nominal and contingency operations. [...] This would aid in acceptance of the new risk environment present in the extraterrestrial environment. 2. Provide both public and private spaces to allow crews some flexibility and choice in how much they are around other crew members. Allowing someone the ability to throttle their social immersion can provide for all varieties of personality types and let the crew social dynamic develop without forcing any out of their mental comfort zone (to a certain extent). 3. Allow a certain amount of reconfiguration of personal and public spaces. Sometimes it is nice to just rearrange the furniture to make a place feel more at home, at least in one's perception. Reconfigurability also can allow one public space to provide for multiple purposes in the community. The galley/dining area that can have its tables pushed to one end of the room or out in the hallway can become the holiday party dance floor but be back to serving breakfast the next morning.

Nice place to work with all the resources. A quiet room to rest and feel happy. Opportunities to explore safely new horizons.

Rest/sleep. Social interaction (and not). Activity /exercise (mental and physical). The ability to adapt should be selected terrestrial by mimicking extraterrestrial habitat.

Variation: the feeling of being able to be in "different" spaces for a fresh perspective, which would prevent participants from feeling cramped or trapped. Greenery: vegetation helps improve mental health (as well as air quality). Social spaces: areas where people can go to enjoy each other's company and take part in bonding activities, to boost wellbeing and group cohesion.

1. A minimum privacy space defined for each resident (same volume/comfort to avoid jealousy) with customization capability of the decoration so that each can choose their preferred appearance (color as a minimum), and with a view of the outside; 2. A gathering room for social time which accommodation includes features specific to the zero or reduced gravity environment, making it specific to space rather than a reminder of Earth places (it is important that whatever design is chosen, it favors the making up of a new mindset of the residents, such that they understand that their space habitat is now their new home, and such that the risk of nostalgia is minimized by avoiding features which could remind them too much their initial Earth habitat); 3. An architecture of spaces which provide clear separation in order to give the impression of being separated habitats and not just spaces within one single habitat. And which would take advantage of the zero or reduced gravity to allow tasks impossible when under earth gravity. On Earth most massive objects lay on the floor or, when smaller, are accessible at body-height. There is no more reason for this limitation in space, on the moon or even on Mars.

Safety—the crew will have to feel safe, to trust that the habitat will be also a shelter. Comfort—the crew will need to "feel at home", will need to "adapt" the living environment to the evolving needs. Personalization—each crew member will be part of a team but will be an individual, with individual preferences regarding the living environment (artificial illumination, temperature, ...), food, relax.

Flexibility, adaptability, adjustability. Did I mention compliance? Conformability? Pick your own synonym!!!

Private space, good food and comfortable place to sleep. Exercise opportunities (non-repetitive if possible). Maximize turnover opportunities (ha!)

1. Personal space (private) for each resident; 2. Worry free electrical power and life support; 3. Safe-haven or emergency return provision.

1. Safety (feeling safe); 2. Adaptable to personal and/group taste/needs; 3. Outside viewing.

6.3.5 What Would You Take?

(Q9) If you were about to go on a long assignment/mission (1 year and more) to an extraterrestrial location what personal items would you take along to make your life better and why?I would personally take more functional elements including e-book reader/electronics, headphones, laptop. I typically do not keep much in the way of personal items around and did not bring photos into my isolation. For longer missions I might bring some photos or other minor adornments for the sleeping berth.

Laptop—so I don't feel disconnected from my regular work routine. Pictures and videos of family and friends. Ready to eat Indian food with good shelf life. Songs and movies that I like—for relaxation and good laugh.

Some type of device (computer or phone or something else) that allows me to store big amount of data with: Pictures and videos of family, music.

Pictures of the loved ones (because they are inspiring!). Good music selection (because I usually listen to music while exercising, cleaning, or performing other "automatic" tasks). Books (because reading is one of my favorite hobbies). Dulce de leche and jamón serrano (because….I just love those! And it would remind me of good time with friends on Earth).

I would bring lots of books and movies to help keep my mind occupied. If possible, I would pack some of my favorite foods (both meals and snacks); food choice seems to take on a larger sense of importance in these environments). I would also bring some hobbies/games to pass the time.

This depends on if there will be gravity or not. I would bring small things to do, to avoid too much boredom—books, games, iPod, knitting supplies, etc.

I'll generalize it to two categories: media to consume and media to create. The former includes music, movies, games, pictures of family & friends & places… perhaps electronic, perhaps physical. They mentally refresh connections with those back home and can serve as conversation topics with those along with you on the mission. The latter allows for creative expression and camaraderie, should the resulting works be shared with crewmates, and the development and maintenance of a cohesive crew culture. A rotation of both consumable and creative media can break up monotony and help crews maintain mental acuity. I also get a lot of mileage out of a comfortable sweater or hoodie.

It's the time between meaningful habitat tasks that determine how you will perform during those tasks. I personally need either an escape or a refreshing challenge to get me ready for the next task. From experience, I was content with a ~900 page book and lots of things to watch. For 45 days, that's all I really needed. For a year or more, I would have to expand that. A hard drive full of books, movies and tv shows. Minimal space, virtually unlimited downtime entertainment. For emotional support, I would probably want a digital photo frame, one that I could put thousands of pictures on. I think that would help me stay focused on getting back to Earth safely. For a third, I would probably bring a pair of drumsticks and a practice pad. I haven't gone a day since I was 8 not banging on something.

Assuming that plants, food, and basic necessities will already be onboard… I would take: Kindle = for leisure enjoyment, mental challenges, or ability to get away. Picture of my kid(s) if I had any = needed for a boost of encouragement during emergency situations. Arts and craft or spare supplies I could tinker with = for leisure enjoyment, mental challenges, or ability to fix/enhance items on spacecraft.

A media device with my music library to help occupy my mind during long periods of exercise. Music is also a great memory jogger, and you can often mentally transport yourself to any point in your life when that song was of significance. This same media device would contain a movie library for downtime or background noise while working, studying or for unwinding after a long day. Pictures from home so that I can "see" my family, even though I won't be with them for a long time. Probably some games that I can play to practice motor skills, keep my brain sharp, or just play with my crewmates in some teambuilding events.

Journal for reflection and recording. Plants and seeds for entertainment and group care. Photographs and videos of people and places at home to maintain connection to life on earth.

I would bring photos and videos of family, books, music, headphones, games and other entertainment for both myself and to share with others. [. . .] Entertainment and headphones for myself since I am an introvert deep down. While I deeply enjoyed my time in HERA and other times I've been with others in close quarters, I often recharge most effectively by myself. I would also bring entertainment to share with others though to be able to develop relationships with my teammates.

About the same as HERA, I guess. A ceramic coffee cup (we ate and drank the same foods as on the ISS. Think "tea out of a metal bag"). My father's issued WWII Bible. I created a GoPro video of riding a bicycle through my hometown for an hour. We would watch it as we peddled the stationary bike. Audio of favorite songs and my son playing the piano. Photos of loved ones. My wife created cards of encouragement to be opened once a week.

Playlist of favorite music for entertainment, encyclopedia for intellectual curiosity, and pictures of loved ones for sentimental nostalgia.

An offline version of Wikipedia. I brought this with me on deployment where we had 0 communication with the outside world for many months at a time, and it essentially simulated the internet. I would read a random article every day and the psychological effect of having access to more information than I could possibly digest during my trip was very positive. My first few [deployments] I didn't have this, and in my (limited) spare time was already calculating how long it would take me to get through the reactor tech manuals we had onboard to read. Something for drawing. Likely an electronic tablet due to paper/weight constraints. Also, a Virtual Reality headset. The games they're coming out with are both immersive and deceptively great exercise now (Beat Saber!), which may help with the psychological lulls that can accompany long programs of repetitive exercise.

I'd take an electronic tablet; which would several of my needs: (1) pictures of my family (2) many, many, many electronic books (3) a method for journaling about the day/experiences (4) ability to track and write the research down in real time (5) electronic correspondence with friends and family.

Books and a ukulele. Those, together with sport (but I guess training equipment would be provided as a non-personal item) have best helped me cope.

Someway to capture photos and edit them, like a cellphone or a camera and a laptop. I love capturing experiences through photography and editing photos helps me unwind and relax and generally makes me happy. Plus, I would assume there would be some way of communicating to Earth and that people would be interested in seeing snippets of daily lives. I really want to say my cat but she hates travel, so maybe a plant. It is rewarding to take care of something and see it thrive. A journal to write down daily memories. Pictures of my family and loved ones. Music—again a cell phone would likely have this feature. Books or a kindle. Always great to have a few movies or TV series on hand to have something mindless to wind down with every now and then. I must say, I do love playing card games or brain games like sudoku so perhaps a deck of cards and a sudoku book. I would also try and bring some games for the crew.

I'd bring a few real books, my iPad for drawing/art, a few non-perishable favorite snacks for the special occasions that might pop up, and perhaps a single weird thing that strikes me in the moment. When I was on my year mission with HI-SEAS, that was a model of the mothership from Homeworld 2, engines that lit up, super dorky and amazing.

Communications & work supporting equipment; cameras; collection of classical music; spare eyeglasses (& maybe wine glasses??); 2 year's supply of medical drugs; family photos; daily routine exercise equipment; work support information; personal backpack.

Photographs of family—maintaining that connection with home. A Kindle or books for down time/entertainment. A pack of cards for group entertainment. Some sort of treat, for example, a small chocolate bar to be enjoyed once a week/month. Also acts as a way of marking the weeks/months.

I would take photos of my family, digital literature with both academic and pleasure reading, and hope to bring a bit of favorite food/snacks to ration out over the year.

Redundant hard drives with all my personal files, music, photos, plus a full copy of Wikipedia/Gutenberg/etc. Sound-cancelling headphones. A few tangible mementos. The best camera money can buy. The best personal computer money can buy. A personal toolkit optimized for fixing/adapting things around the habitat.

Electronic media: books, movies, photos, music for entertainment, work and distraction. Camera. Some favorite items from earth: t-shirt, blanket. Coffee and wine.

My personal computer for personal work, internet connections (even if no connection, I could still do a lot a science offline), and relaxation (watching movies and listening to music).

1. A photo frame with my family (electronic so I could have different ones every day). 2. My music playlist and podcasts. 3. My bonsai tree if I can.

My grandmother's necklace, picture and music. Mostly digital files.

Some favorite clothes. A few posters (assuming I can put them up). I would definitely need some means for regular exercise. Games to play with the other crew. Special indulgences to hide away for birthdays and other special occasions.

A variety of ways to record and explore the experience in real time, musical instruments, art etc. The experience of ICE for over a year must be physiologically digested in real time, a person can't just carry the accumulating baggage. This is why effective ground support is critical.

(1) A computer with all my private/personal files and programs on it (entertainment materials to study materials, etc.). (2) An assortment of storable food items that other team members might not have or know about. In reality, 1 year is very short and goes by very quickly. And most likely people will think they need more than they really do. Now if it was open ended, unknown return, this would be a very different question. An interesting personality observation/question here is with regards to people who can and can't eat the same thing every day for a week, who can't wear the same clothes multiple times in a row, and such.

Something from home: images, drinks/food/stuffed animals from family; letters to be opened mid-mission—anything that connects me to Earth. You wouldn't believe how a poster of a green forest and a crystal blue lake in a personal space can be powerful reminding you where you come from and where to you'll return some day. Some old-school explorer items: I always bring a compass (although we have much more modern tools), a self-wielded knife and my old-school cowboy-hat (since many missions—a genuine Australian Stetson)—to connect to the previous ages of explorers. We even take a logbook-print and business card from a historic Austrian polar expedition from more than a century ago with us. One item stays at the Mission Support Center, the other goes along into the desert. A ridiculous colorful Hawaii-shirt and stupid looking clothing to allow for letting steam by means of fashion.

I'd take a good stock of digital photos of family, friends, and places there were not going to be part of my life for the duration of the mission. Visual reminders of other people and places are great supports when you cannot easily interact with or visit them in person. I'd also take along plenty of reading materials regarding a wide variety of topics. Being able to immerse oneself in something outside the new daily norm can help to alleviate stress or any tendency to allow one's perceptions and mindset to narrow, given the likely constrained living conditions and inability to casually physically leave and experience something different.

At least ten (10) good books. For once a week reading.—My favorite snacks: chocolates, biscuits, dry fruits and seeds. [. . .]. Five preferred liqueurs, at least one small celebration every 7 weeks!—Music and movies. . .no limits, in digital format.

Books (both for learning and entertainment). Audio equipment with music/ podcasts etc. Exercise equipment. (Keep the body in shape will affect mental stability positively).

Download probably all of Youtube before you go!

Yoga or other exercise kit—it's easy to forget that you need to move, when your daily routine consists of moving around a space just a few tens of meters across. It would sometimes feel like a drag to go along to yoga, but the moment I would begin I would feel instantly calmer and happier—even if I didn't realize I needed it. Chocolate—comfort food is important! I'm not sure how I would bring a year's worth of chocolate with me though.

I have the impression in my personal case that I don't need much to make my life better, so I will simply answer based on my own space flight experience:—photograph of the family—some personal music and poetry which had some inspirational impact in my life,—some puzzles which could entertain my mental skills in a different way than the mission goals themselves at work (I took the Rubic's cube with me on my 3 flights)—some fundamental physics and philosophy books which would help maintaining high level thinking/reasoning/questioning about our human condition—all sort of games to play with other crew members during part of the free time which would maintain social (and hopefully safe) interaction like card games, chess, sport,

Pictures and voices of my family, music, books.

J.S. Bach; a digital reader; an amateur radio station (or similar off-line, off-the-record communications capability). Music and reading are great healers and distractors. The same is true for being able to communicate with others external to the intense and controlled mission.

Digital music/books; Comfort clothing (comfort food would soon run out); Games?

1. Bible. 2. Journal. 3. Photos of family. 4. Sketch pad/device.

In addition to communication devices, I would take tools for painting and sculpture.

6.3.6 Other Important Aspects Related to Habitability

(Q10) Please feel free to comment on any aspect of habitability that you feel is important to take into consideration for designing effective extraterrestrial habitats for the future.

I noticed that the lack of crew conflict we experienced had some basis in adaptability to seemingly trivial details such as layout of the storage space not promoting conflict among crew members who may have different approaches to stowage, etc. Minimize collisions of personal style in designing the stowage and common areas which can highlight personality differences and approaches. In having very little space to work with, our crew had limited options for setting up things like stowage and personal effects, and that lack of personal input likely suppressed conflict. Designs should allow personal freedom in berths and personal spaces, while emphasizing utilitarian design with limited choice in common areas.

To be able to communicate with mission control and loved ones (effectively and frequently); availability to watch news, sporting events, etc. in real-time/live to stay connected with world affairs. To be able to share your experience back home with kids to inspire the next generation. To get regular updates on how your work is getting traction back on Earth, and how everyone is excited about your exploration and value your work.

Human capabilities and limitations considered. Adaptable.

I personally think that trying to avoid the feeling of being "confined" in the habitat is gonna [sic] be very important for extremely long duration missions. Having habitats that allow us to see the exterior would have a very positive impact. Or maybe, not necessarily having "transparent" habitats, but walls that can (virtually) give the illusion of being in an open space.

As noted earlier, I think providing different areas of the habitat with barriers between them would be very helpful (such as the two different levels of HERA). Also, a private space for each crew member would be important over the long term.

I loved the sleeping pods in HERA—very cocoon-like, cozy and quiet. These are a good design for sleeping. It was difficult for our HERA crew to all site around a single table because all the tables were against the wall. It would be very nice to have a table that all crewmembers could sit around and face each other, rather than face the same wall. Lights and sound systems should have an option for voice-

activation. This seems standard in many homes now and would be a nice feature in a high-tech extraterrestrial habitat.

Reduce the amount of manual effort required to keep the toilets working. Keep working toward closing the loop on life support systems, and on keeping the inside and outside separated in the most meaningful ways (i.e., don't track regolith into the hab from the airlock). Provide occupants with optionality for use of the hab volumes to the fullest extent possible within the SWaP constraints (size, weight, and power). People are highly adaptable, but processes for using a volume should be as simple and straightforward as possible.

When you are part of a mission, you are pretty focused on the task at hand. Your motivation for why you are there never wavers too much. Under that mindset, you never really think "why was this built like this" or "you know, this could have been better if they put this here." You adapt to what you are given. So, maybe at the end of the day, you only need so much out of your habitat living environment. With that said, maybe over a year, those thoughts and feelings would tarnish, and you would like a more comfortable environment to live in.

Minimizing light & noise pollution. Fragrant deterrents should also be considered. Our senses are important, and there is a lot of research on how these three senses affect us. Light and noise pollution engineers should be involved when designing these habitats.

A mechanism for crew privacy in their CQ's is important. The ability to contact family if/when needed should be provided. Adequate stowage locations for provisions and science should be provided as well. Often on ISS, we run into the issue where we have so much stuff flying up on visiting vehicles that we have trouble finding places for it to go. This can really impact the crew day, mood, and efficiency when they have to move bags out of their way to do daily tasks. Flying exactly what is needed to keep the crew fed and hydrated, having some emergency provisions available in the event of a landing waive off, and keeping open space for experiments that will take place on the mission is going to be critical going forward. For long duration missions to Mars or beyond, this isn't likely to be as big an issue, but lessons learned from ISS should be incorporated going forward in future design.

Some practical amount of detail should be spent on aesthetics. A cluttered space can be less conducive to productivity, but a totally bland space can impact one's state of mind as well. There's a reason our homes are covered in decorations and not just blank walls.

*As humans, we have a connection to earth's environment that I was not aware of prior to my HERA experience. A place that is not 100% predictable and that changes as flowers and plants (and perhaps animals) grow will help. The lack of four seasons or even weather at all can be monotonous. Designing lights, sounds, and other aspects that are random and/or change with seasons and holidays could be beneficial. Also, one place to escape to with deep connections to home and earth will enhance crew well-*being.

External views either real or simulated, both shared and personal spaces, separate work and personal quarters for sleep hygiene.

Privacy in the bathroom. On the sub it was pretty straightforward, though the bathroom was very small. In HERA, we had a privacy study being run which removed the door and had only a curtain around the toilet. Interestingly, there was a white noise generator next to the toilet which made a huge difference psychologically. Even if you can't have a hard barrier there, the audible privacy has a large psychological effect and doesn't have to be particularly intrusive. Some public bathrooms do this too with music.

For me, the personal space, shared space and windows are the key. Along with the belief that the habitat is safe and well-built and will keep me safe during the duration of the mission.

Spontaneously, I would say that one should aim (aside from radiation protection and other technical considerations) for a warm, ergonomic, and easy-to-clean environment.

WINDOWS! A space for socializing, a space for personal isolation, and a space for working out.

People have needs that aren't mission related. I realize space is tight and it's expensive to ship stuff, but we're going to have to treat future long duration habitats as places that people live, not just work. The importance of being able to indulge in all the facets of human life, including ones that aren't based solely on work/science productivity, will be very important. Missions fail if the crew loses it just as surely as they fail if the doors all pop off or the reactor blows up.

Designing to the environment is crucial—Pressure/humidity/radiation shielding/dust storm protection. Also need to adapt to the users. Adequate space for the numbers, appropriate size and height for users. Consider the cultures of users if needed.

The designers must have first-hand experience living and working in a simulated habitat. And even then, the habitats they design must be adaptable/configurable by the crew during the mission because—at least for the first few decades—we won't really fully know what works and what doesn't, habitability wise.

Redundant life support systems. Individual private spaces. Outside view. A "fun" area. Climate control.

Do not forget to include a sports facility, e.g., a treadmill and a small sauna (so important to relax and to talk informally with other people).

Communication system that facilitates conversations with earth/people outside. The time delay means that these conversations will be long, so need facilitation so they can be weaved in through the workday. Facilitating external psychological support is critical but takes time. People who don't think they need external psychological support don't understand what they are facing. Long duration ICS simulated missions in functionally similar habitats are critical for training and habitat testing.

a) Efficiency and safety: short distances between related tasks; high situational awareness on station health status (that also includes windows to monitor conditions and people outside). Back-ups, redundancies (what is a crew quarter module is risking depressurization, where do people sleep?) You need to be able to point at ANY element of the design and ask: what will we do if that thing fails? b) One size does not fit all: depending on mission phase, the usage patterns evolve, both because of some experiments are finished, or/and personal preferences evolve etc.... so the more modularity you have, the more you can adapt to current needs. Do not assume to know the personal preferences as the mission evolves—sometimes we as analogue astronauts don't know that very well in advance.

I think at the core of the design, the reliability and trust the crew can have in their habitat will go a long way to helping all aspects of the mission. If it feels safe and secure, it can feel like home, at least for an extended temporary time frame. And if it feels like home, people will fare better and get more out of the experience. Even though South Pole Station is a hazardous and isolated location, I still consider it a former home and miss it on a daily basis.

This is more about social structures than physical spaces—but I firmly believe that having a diverse leadership of any extraterrestrial base is essential. When a bunch of white males are in sole control, however well-meaning, I worry that things would quickly get regressive. Spaces that encourage a flat hierarchy would also be beneficial to group wellbeing. Having leaders separated in any way from the rest of the group could lead to feelings of division and would risk creating tensions in the group dynamic.

If I were going on a long voyage in space, I would like to find an environment (habitat) which would help me feel at home, but not a mimic of my earth home but on the contrary a totally new kind of home, which would be totally different from any kind of home I have been used to when living on Earth. A bit like when you go on a trip to a new (unknown to you) place on Earth, and you find it so nice and compatible with your own personality that you decide that you would easily come and live the rest of your life there. In this respect, the habitat must provide attractivity, easiness of life, a view, a diversity of spaces composing the habitat, and people you meet there that are nice. Even if the habitat is perfect, if there is incompatibility between crewmembers, life will still be a pain. There comes a huge responsibility of the crew selection process.

Flexibility, adaptability, variety, and accommodation for improvisation must be given reasonable weight alongside the more traditional system trade-offs like safety, energy, etc.

There are many ways to design extraterrestrial habitats. Concepts placed in the distant future are difficult to assess because of the lack of constraints. A more profitable approach is to look at early habitats using today's technologies and for your book, avoid a single point design. In fact, a good model for your results would be the space equivalent of Pattern Language by Christopher Alexander. Point designs have a very short life, but an extraterrestrial Pattern Language could be influential for decades.

I think it's very important to respect every crew member personality and give them possibilities to create their own environment where it's possible.

Reference

Wordcloud.co.uk, https://www.wordclouds.co.uk/

Chapter 7
Looking Forward: How to Convert a Tin Can into a Home: Exploring Solutions to Selected Dimensions of (Socio-Spatial) Habitability

This chapter focuses on a selection of major overarching habitability issues which have been identified by experience and research to date. Each factor is discussed and commonalities and differences identified to enable both synergy and effective distinctions where needed. For each, we explain the underlying socio-spatial relationship (the context) with reference to latest research followed by examples from isolated and confined environments with possible solutions for each phenomenon.

Among the myriad future possible solutions, some will address multiple needs and not all solutions can be applied in all instances. However, to use this large 'pool of ideas', mission planners, aerospace architects and designers have to understand the underlying human-spatial concepts, which are discussed for each habitability issue. Only then, will it be possible to come up with a solution that deserves the name habitability design solution. Habitability issues discussed in this chapter encompass the reciprocal conditions of the individual and small groups with mission and environmental characteristics (place and space), lengths (short- and long term), resources and human activities within an extraterrestrial habitat.

7.1 Introduction: Can We Build THE Perfect Habitat?

With this book, we outline the factors that accumulating evidence strongly points to as critically important to building functional, supportive, persistent long-term habitats for extraterrestrial locations. It should be apparent by now that existing habitats in isolated, confined environments, both extraterrestrial and terrestrial, with few exceptions, are rather impoverished places to live in (see Chaps. 4 and 5). The reasons are easily understood. First, these environments are characterized by significant challenges in both logistics and environmental demands; and second, current design strategies emphasize meeting minimal requirements with the least resources necessary. Thus, the focus is still on building functional survival shelters and not living places.

S. Häuplik-Meusburger, S. Bishop, *Space Habitats and Habitability*, Space and Society, https://doi.org/10.1007/978-3-030-69740-2_7

Over the history of space habitation, we can say that **the more transient the occupants, the less emphasis or concern is spent on human comfort**. This has been true as to what occupants have been willing to accept and the degree of time and effort designers have been willing to invest. As soon as the perspective shifts to answering the question of what is needed to *thrive*, the list of necessities changes. While less-than-ideal conditions are bearable or even desirable for short durations, '*camping*' is not a viable long-term approach (see Sect. 7.4).

To complicate matters further, in most extreme environments many of Earth's inherent natural conditions must be provided through technology, e.g., breathable atmosphere, shirt-sleeve environments, day/night cycles, processed foods, etc. (see Sect. 7.6). Bionomics, the study of the relationship of organisms to their environment, would argue that these substitutions may not be viable alternatives for the entirety of conditions that they seek to mimic. For instance, *does a picture of a forest on your wall adequately compensate for never interacting with actual plants?* Therefore, it is critical to identify the following:

1. Are there factors related to Earth's natural conditions that are needed/vital for human wellbeing in ET and EE Habitats?
2. If so, what are the characteristics and dimensions of those that need to be surrogated?
3. And which technologies and socio-spatial strategies are available to do so?

Thus, it is apparent that the human is the most fragile and variable element in the system (Fig. 7.1). Hence, there is no such thing as 'THE perfect habitat', fitting all inhabitants and every situation (Box 7.1).

Box 7.1 One does not fit all

The problem 'Why we haven't yet built THE perfect habitat (that fits all and every situation)' is manifold and can be summarized as follows:

1. What we know is based on <u>imperfect terrestrial analogues</u> and a few orbital/transportation vehicles—all of which involve limited number of crew members and limited duration. Thus, our ability to generalize critical factors for a permanent ICE habitat intended to be occupied for extended durations (years) is almost nonexistent.
2. <u>Only a few people</u> that can report on outer-sphere experiences are available. All of them come from a small unrepresentative segment of human beings (e.g., well-educated, mainly male adults, little geographic or cultural diversity, etc.)
3. There are <u>few existing opportunities</u> to test new habitat designs in locations that are located in remote locations here on Earth (aka analogues) that are not already well established (aka settlements in polar regions or remote island chains).
4. All our perceived experiences are deeply entrenched in cultural practices. <u>Cultural mindsets</u> are powerful lenses that shape the desirability/

(continued)

Box 7.1 (continued)

acceptability of all things, including what we expect in our living spaces. Many of the existing analogues involve homogeneous participants from particular national or cultural groups. Thus, generalizability is severely restricted. What is acceptable for one group may be completely unacceptable to another.

5. There are both individual and group processes that must be addressed in designing functional, supportive facilities. These foci are highly interdependent but different. Fortunately, there also appears to be certain types of personalities that function best under the kinds of conditions found in ICE environments so we don't have to design for the entire gamut of human personality.
6. All the above mentioned are in interdependent and dynamically changing relationships.

So how do we identify, design, and finally build living spaces that will address wellbeing and function at both the individual and group level that are not dependent upon any particular cultural or personality perspective?

The authors' individual and joint research of 30+ years is based upon the conviction that a very strong and interrelated relationship exists between humans and the space we live in (Fig. 7.1).

Humans, as an individual or group, are in constant interaction with the physical and social environment. The diagram in Fig. 7.1 highlights the dynamic process of social life. There have been many theories built on this relationship. This illustration combines two different approaches: (1) the 'Engineering control theory' proposes that a trigger, i.e., an event or situation, causes a chosen human action. Triggers can

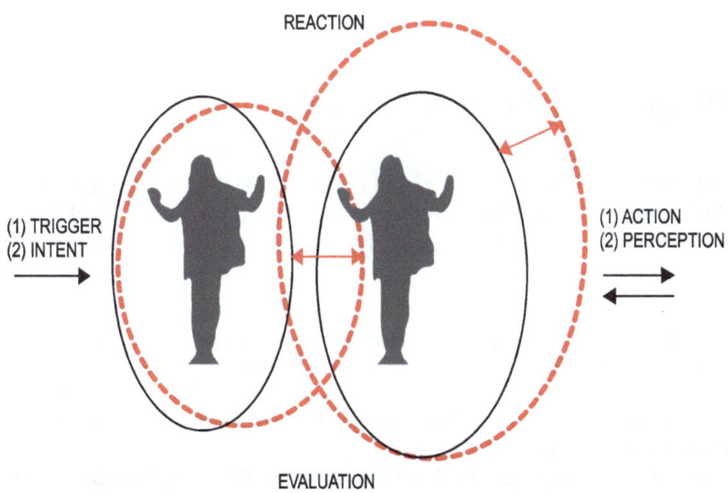

Fig. 7.1 The versatile system between humans and space (Credit: The Authors)

be external, internal or even synthetic. Resulting actions cause—social and spatial—reactions. A newer (2) 'Perceptual control theory' has experimentally shown that individuals appear to control their perceptions, rather than actions.

Although there are multiple efforts to explain this interconnected system through various scientific and mathematical means, this socio-spatial relation is not yet fully understood and cannot be calculated. However, there are two facts of which the authors of this book are certain:

1. There are basic human commonalities to adaptation to the kinds of stresses inherent in extreme environments *(trigger—action/intent—perception)* that are characterized by isolation, confinement and extreme environmental risk (ICEs) and
2. Human adaptation to these environments (*reaction—evaluation*) is affected by the living spaces in which the inhabitants live and work (either supportive or limiting).

Among the myriad factors that will compose the future solution(s), there are those we believe are critical. Some are obvious ... some not so obvious. Some solutions will address multiple needs and not all solutions can be applied in all instances, e.g., what may work for a planetary base may not be feasible for a transit vehicle. In addition, while the exemplary design suggestions will be based on current imminent or existing technology, with the rate of technological innovation and invention, some of these will have already matured into practice by the time this book is distributed and new possibilities will have emerged. The intent is to get us all thinking, not only *outside the box*, but outside the confines of our terrestrial mindset. With the current rapid development of technology in mind, architect Cedric Price's famous cautionary question remains pertinent: "*Technology is the answer, but what was the question?*"[1] For it is the human being that is the driving force for and behind all decisions.

Let's move on, we have a universe to settle.

7.2 Designing for the 'Best Fit' Person

Originally, a great deal of time, effort and research was put into trying to identify the 'right stuff',[2] i.e., the best personality for space. It soon became evident that the kind of person that was exemplary for the early space programs (i.e., military test pilots

---

[1]In 1966, the British architect Cedric Price titled his lecture: "*Technology is the answer, but what was the question?*". He was an architect constantly searching, questioning and rethinking and invited the audience at his lecture to search, question, and reconsider the impact of technological progress on architecture. https://www.pidgeondigital.com/talks/technology-is-the-answer-but-what-was-the-question-/chapters/.

[2]'The Right Stuff' is a book written by Tom Wolfe in 1979. "*What is it, I wondered, that makes a man willing to sit up on top of an enormous Roman candle, such as a Redstone Atlas, Titan, or Saturn rocket, and wait for someone to light the fuse*? (Wolfe 1979, p. 2).

with high tolerance for risk and life-threatening challenges) were not well suited for the more mundane tasks involved with routine shuttle missions (i.e., task-focused, goal oriented, operationally minded, high focus on risk avoidance, high tolerance for repetitive tasks). Even more eye-opening was the discovery that some of those individuals who performed well in Shuttle missions were not particularly well suited for in-situ, residential mission such as MIR or ISS. So, in addition to looking for critical differences in the very small pool of individuals who had actually gone into space, researchers went looking at the larger population of terrestrial analogue crews to help explain what was anecdotally observed in space crews.

7.2.1 Ahh. . .Mission + Personality!

Early evidence pointed to differences in mission demands as a critical difference. The kind of personality that was best fit for testing new technology that pushed the frontiers of flight (pioneering) was distinctly different than the mission profile for crews tasked with piloting the shuttle to ISS and back (expeditionary). And the mission demands for tolerating a long duration stay in an isolated, confined habitat (residential) were distinctly different from those of the test pilot or the Shuttle crew (transitory). However, it wasn't just that the mission profiles demanded different roles; it soon became clear that *different* personalities were *better fits* for those roles.

For instance, evidence for the expeditionary profile came from studies on those that did well in successful polar trekking groups and found that individuals who were loners seem to be supportive of better adaptation to the expeditionary mission demands, i.e., they were happier alone than dependent on others, highly autonomous, independent, uncomfortable about and relatively uninterested in accommodating others in a group, task-oriented and somewhat competitive (Rosnet et al. 2000). It was understandable that the intense task focus of a polar trek, in which each individual is highly autonomous and individually self-reliant during the long travel each day, situated in an environment that precludes group interaction except for fundamental coordination of locomotion across the terrain, *selects for* individuals that are distinctly different from those who would occupy a habitat or a confined environment for long durations. In other words, only individuals with this inward, self-focused personality would find such challenges rewarding and be successful at these tasks.

Similarly, a different adaptive personality profile emerged from studies of polar winter-overers that is characterized by low levels of neuroticism,[3] desire for affection, and need for order, as well as a high tolerance for lack of achievement or boredom, which would fit well in an environment where isolation and confinement

[3]Neuroticism, in psychology and development, a broad personality trait dimension representing the degree to which a person experiences the world as distressing, threatening, and unsafe. Each individual can be positioned somewhere on this personality dimension between extreme poles: perfect emotional stability versus complete emotional chaos (Kwon and Weed 2019).

prevent accomplishments, lack variety or stimulation and the participants experience frequent shortages and problems (Palinkas et al. 2000). Those that would best adapt would be those who could more quickly adjust their expectations to the immediate situation and tolerate such obstacles. Neither of these profiles were even remotely similar to the 'right stuff' profile of early space crews largely composed of military test pilots!

7.2.2 PLUS the EXPERIENCE of the Environment!!

This realization that the *fit* between mission demands and personality type was critical was further coupled with studies that produced supporting evidence for a focus on the situational or environmental *experience* as the defining factor rather than the particular environment per se.

For instance, being the first crew for a mission to Mars will entail elements of pioneering and pushing the frontier but the experience over the majority of the mission will be one of in-situ residency. This comprises living with a few others in a small, confined and severely limited habitat, both during transfer in the spaceship, and after landing in a Martian habitat. A typical 'test pilot' or expeditionary personality would be significantly distressed by the huge quotient of monotony, lack of stimulation and restriction in activity. The challenges of 'living together' will be similar for all long-term missions. Even for lunar and orbital missions, where the trip will not be as prolonged, the terrain less challenging and better known than Mars, the confinement and its implications for living will be no less.

7.2.3 Summary

Psychological research to date seems to support two general findings (Bishop 2011):

1. There do seem to be consistencies in the personality profile of functional and dysfunctional teams across environments that produce similar experiences, and
2. Characteristics of the mission may define and necessitate very different personalities.

Thus, instead of trying to figure out what the single best personality would be for space missions, it seems it is more important to select the 'best fit' person and team. The characteristics of mission and environmental demands and the expected experience (e.g., long versus short duration, station-keeping versus expedition profiles) will provide better selection criteria for hardier and better-fit personalities and teams.

7.3 What You Take Is What You Have: Limited Space, Limited Resources and Limited People!

Seems obvious, right? Yet anyone that has taken a vacation, gone on a business trip, or traveled will attest to the *near certainty* that you will leave something needed behind. Doing so for a space mission means largely doing without or making do with some kind of substitute. Looking forward, new computer-aided design and manufacturing technology (such as 3D printing and sintering) offers intriguing possibilities for replacing a vast array of needed components but that can only happen if the required raw materials are available. In addition, new technologies always produce a number of challenges not previously seen (such as recycling, reuse and materials management). One unbendable fact about the logistics of space travel and habitation therein is that ***everything is limited***; physical space, storage, air, water, food, power, fuel, medicine ...everything, even people. Once established, possibilities to set up in-situ or remote mining and resource acquisition will be explored. But until that capability is in place and the inventory of accessible resources identified, we must plan to make do with what we pack and bring. Such finite resources mean building to thrive just got a lot harder.

7.3.1 The Question of 'Adequate Volume'

When designing a habitat, available space and its allocation is one of the basic parameters. This is true on Earth, and even more for the design of an extraterrestrial habitat. The principle '*form (volume) follows function*'[4] dates back to the early twentieth century, and early reference books for spatial requirements in building design were published in the 1940s (e.g., Neufert Architects' Data). Within the industrial arena it became tempting to having a guideline for the production of goods and also space. For extraterrestrial habitats, the 1987 NASA-STD-3000 standards (NASA 1978)[5] set a guideline for the design and sizing of space modules. It was created to provide a single, comprehensive document defining all generic requirements for space facilities and related equipment which directly interface with crewmembers.

Those, and similar reference guides, are based on ergonomic design principles and include precise drawings and measurements for a number of activities. The basis for those drawings are the metrics of the human body—size, measurements, range and body posture considerations (Fig. 7.2a, b). As illustrated, the 'average person' in

[4]The phrase "form follows function" was coined by architect Louis H. Sullivan in his 1896 essay "The Tall Office Building Artistically Considered." The statement refers to the idea that a skyscraper's exterior design should reflect the different interior functions (Craven 2019).

[5]NASA-STD-3000, the Man-System Integration Standards revised in 1995 as NASA Std-3000B; and in 2010 as Human Integration Design Handbook (HIDH) and NASA-STD-3001 Space Flight Human System Standards Volumes 1 (Crew Health) and Volume 2 (Human Factors, Habitability and Environmental Health).

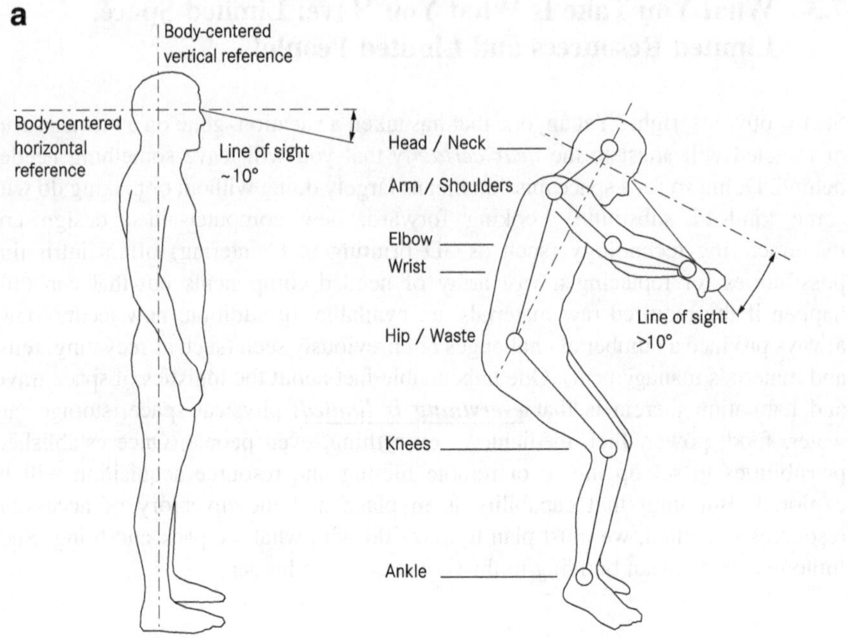

Fig. 7.2 (**a**) Figure comparison between the neutral body orientation on Earth and the one in microgravity. The head is tilted down; ankle, knee and hip height increase; elbow, wrist and shoulder are raised and elbows are abducted (NASA [HIDH] 2010, p. 49). However angles vary according to the individual. (Credit: as published in Häuplik-Meusburger 2011, p. 19). (**b**) Comparison of the neutral body position depending on different gravity conditions (NASA, Drawing by Brand Griffin (Griffin 1978))

terms of anthropometry and biomechanics is defined through a specialized vocabulary. Designing and sizing of space modules is based on the anthropometric models of the *90-Percentile*, which theoretically provides coverage for 90% of the user population. For habitats in a microgravity environment anthropometric design is extended due to the change of the Neutral Body Position[6] As illustrated in Fig. 7.2b there is a clear difference from the neutral body position in space compared to Earth. At that time, it was believed, that "*there is a definable relaxed body posture in zero-g and that the eligible flight crew population can be fitted to that posture*" (NASA [Bull.17] 1975 p. 2). Today, we know that the range between those positions vary much more. However, there are some similarities that have large consequences for the design of the whole space station interior: **The head is tilted down, the arms and legs lift up, and there is no pooling of fluids in the lower extremities.**

Over time, many studies have tried to ascertain an exact number to answer the question: "*What would be an adequate space for an inhabitant?*" In 1999 space architect Constance Adams clarified the term 'habitable volume' as total free volume available for human habitation and use (Adams and McCurdy 1999). In 2008, Marc M. Cohen proposed that pressurized volume increases as a function of mission duration, both as a power curve and a logarithmic curve. He also showed that this volume trend does not level off but continues to rise (Cohen 2009). Table 7.1 shows a comparison of historical space stations, and how the absolute volume as well as the habitable volume has increased over the years. Despite those efforts, there is yet to be a consensus reached on a specific m^2/m^3 per person number. Challenges for the calculation of appropriate habitable space is driven by the fact that different (mission) scenarios, people and environments impose different spatial requirements.

Box 7.2 Comparing Living Space

When looking at the numbers of actual and suggested m^2 per person around the globe for terrestrial habitats, volumes proposed vary widely. In addition, the fact that different variables are considered critical render comparisons impossible. For example, the Japanese are acknowledged masters of making the most of limited living space with only 4.8% of land utilized for residential purposes for a population of 127 million. According to a 2018 nationwide survey carried out by the Association of Real Estate Agents of Japan (realestatejapan.Blog. 2018) a single person living by himself in a city center or suburb *should* have a *minimum* of 25 m^2 (270 ft^2) of residential space but the *ideal* amount of space for a single person living a city is much higher: 40 m^2 (431 ft^2). However, living in the countryside raises the guidelines even higher to 55 m^2 (592 ft^2), more than double the minimum recommendation for a city-dweller. In comparison, the average floor area per person in the EU is 42 m^2 (458 ft^2), with the difference between the highest (Cyprus) and the

(continued)

[6]The neutral body posture is when all muscles are in their neutral state and require no extra energy.

Table 7.1 Habitation volume, duration and crew of historical space habitats and the International Space Station

Habitat/ Space Station	Habitable Volume (m^3)	Volume per person (m^3)	Mission duration (days)	Crew
Apollo CM (with LM)	10 (6)	2.22 LM: 4.27	6d 3 h (Apollo 8)	3 (2)
Skylab	320 (OWS: 270 m^3)	120	28–84	3
Salyut 7	90	33–55	16–237	3
Mir	380 (core module: 90 m^3)	45–181	73–438; Longest space mission by Valery Polyakov with 438 days	Resident crew of 2, plus visiting crews, later up to 6 (not including shuttle missions)
ISS	935 (at assembly)	85–201	Varies, 180–340; Longest space mission by a woman: Christina Koch with 289 days	Initial crew of 3, since 2009 nominal permanent crew of 6, plus visiting crew

Sources: Cohen (2009) and Häuplik-Meusburger (2011)

Box 7.2 (continued)

lowest (Romania) being about 45 m^2 (484 ft^2) per person (EC 2020)! In Africa, people have less than 20 m^2 (215 ft^2), even less than the Japanese city dweller (UN Population Division 2000). According to the Organization for Economic Cooperation and Development (OECD) (2020), Americans enjoy the most space per person—an average of 61 m^2 (656 ft^2), with Australians reporting an average of 51 m^2 (549 ft^2) of space at their disposal. Compared to the average living space per person, the U.S. area is nearly twice the size of what a Brazilian resident enjoys at 32 m^2 (348 ft^2), and 19 m^2 (202 ft^2) more than what the British residents have at 42 m^2 (454 ft^2). Compared to the 25 m^2 (269 ft^2) minimum for Japanese city dwellers, risks for confinement stress in future extraterrestrial habitats will exist for most peoples of the world!

In addition to the cultural and location factors, the fact that all activities in extraterrestrial habitats will take place in a reduced or microgravity environment must also be taken into account. In an orbital station or transfer habitat, living space is measured in volume [m^3] since all space (up, down, sideways) is usable. Whereas, in a reduced gravity environment, e.g., Lunar or Mars, physics will once again impose a true up/down orientation. Even microgravity stations are expected to implement some kind of artificial gravity for living areas. Thus, the demand for 'adequate' volume will once again come down to adequate m^2/ft^2 per person subject to all the cultural, physiological and individual preferences.

Such factors eventually made it patently clear that specific recommendations for 'adequate volumes' can only work as guidelines. Following the 2011 Habitable Volume Workshop (Simon et al. 2011), the research team stated that *"the volume required for any duration is a function of multiple factors that contribute to the adequacy of a volume, such as how the interior is laid out and how spacious an interior is perceived."* (Simon et al., p. 2) This has not been a new discovery, a similar conclusion was already stated in the 1985 HuSH study (Wise and Rosenberg 1986): "It's not how large you make it, it's how you make it large!".

Characterizing these multiple factors has yet to be fully accomplished. Possible integrative solutions on how to address the layout of the interior and its perception are given, for example, in Sects. 7.6.2 and 7.7.4. Built and inhabited results of 'adequate volume' estimations can only be as good as the estimations on expected human activity in this spatial envelope.

7.3.2 Limited Space = Limited Availability!

McMurdo Station, Antarctica, exists on the only continent where few resources grow naturally. As with future initial extraterrestrial bases, everything needed for living at McMurdo needs to be shipped in on cargo ships and periodic restocking flights by DC10 and LC-130 Hercules cargo planes. The amount of supplies for a large facility is staggering. Over 12 million liters of fuel (3.2 million gallons) and tons of supplies are brought in every year. Outgoing, 873,120 kg (1,824,900 lbs) of non-hazardous solid waste and approximately 230,920 kg (515,710 lbs) of hazardous waste is generated and removed annually. The station must have approximately 63,200 m^2 (680,000 ft^2) of dedicated storage space for materials and supplies in 22 buildings (National Science Foundation 2019). Figure 7.3 shows just a small section of McMurdo's supply and storage area. Imagine the logistics of transporting and storing this much supplies and materials for a base off-Earth.

As a training ground for extraterrestrial habitats, Antarctica's isolation imposes a reality check on the presumption of plenty employed by the bulk of humanity. Yet compared to what will be required to supply facilities off-Earth, provisioning McMurdo is a walk in the park. Clearly, we cannot approach extraterrestrial habitations with the same mindset of 'availability' we have with even the most remote and challenging locations on Earth.

Actual space in terms of area and volume is valuable in Antarctica, and even more so in extraterrestrial habitats. It will be essential to think how space is used over time. For example, on the drawing board previous space modules tended to look spacious (cf. Fig. 3.2a, b concepts for Shuttle and Mir space stations). However, after a few years of usage, all of them have ended up as cramped and often messy spaces. Things add up in a confined environment over time, including onboard the Mir space station

Fig. 7.3 A supply storage area at McMurdo Station in 2014 (Credit: Elizabeth Mockbee, USAP, NSF)

as well as the International Space Station. Usable and precious (expensive) habitable space turns into storage space, leaving less space for work and living.

Furthermore, in addition to the actual storing of goods, space that is needed for handling incoming and outgoing goods is often underestimated (Box 7.3). Available 'empty' storage space needs to be provided in times of increased demands and integrated into the planning. For example, when resupplies are delivered, a lot of objects have to be transferred from one place to another and temporary storage demands increase (cf. Häuplik-Meusburger 2011, p. 194).

Box 7.3 Managing Objects

The number one challenge by far—in terms of living in space—is just managing objects. Where are things? And where to put things? Continuously. [. . .] And in space we manipulate hundreds of objects every day. And keeping continuous awareness of where they are is a challenge. And when you use them, when you are done with them, where to put them? [. . .] Especially when you resupply the Space Station with the Space Shuttle, it's like you have two houses, side by side, one is full of new resources, new furniture, but the other one is full with old stuff, that you have to get rid of. So you have to start to make some space somewhere, in the old house—The Space Station—so that you can start to bring some new stuff in and free some space in the Resupply ship that will be used for waste. And resupplying the space station is a challenge for that, because you have to think about 1000s of objects, which one goes where, when? The resupply process takes about 1 week, when it's a resupply ship. When the resupply ship is capable—technically—to stay for a long period of time, like the Space Ship Jules Verne, . . . it docks and the crew goes in it, only when they need and the resupply process is spread over the 5–6 months of stay. This way, after

(continued)

Box 7.3 (continued)

about 5–6 months, all the resources have been used, and overtime it has been filled with waste. And then it's easier. It's not as stressful, when you have a short time to do re-supply. Then the crew has to empty everything inside and fill it with waste or with things that have to go back to the ground within a few days.

Source: From a personal interview with Jean-Francois Clervoy in 2009 by Sandra Häuplik-Meusburger (Clervoy 2009, manuscript unpublished)

7.3.3 Microgravity = New Spaces

In the early days of space exploration, the capsules were so small that there was a maximum allowable height for the first astronauts.[7] John Glenn, one of the Mercury Seven pilots famously remarked on the snug fit of the Mercury spacecraft, "*You don't get into it, you put it on*". "*Inside it, you feel like one more piece of equipment—the most important piece, of course*" (Schanche 1959, p. 112).

Today's spacecraft systems have become larger, but, until we have the technology ready to build in situ, the habitat size and mass is still dependent upon the transportation system used. The combined constraints of engineering (what is possible) and economics (what is affordable) result in a severe limitation of the available living space. While similar on-site mass and volume restrictions apply to almost all ICE environments (e.g., polar research stations), they are more complex and, henceforth more demanding, for extraterrestrial habitats, as everything must get off the planet with the requisite additional burden in technology, logistics, force and energy.

Solutions to these limitations have taken various approaches to maximize availability. The overall architecture of the International Space Station uses a **modular architecture approach** in order to create a large space station complex. The modules used, and, thus, their limitations on mass and volume, were developed to fit available transportation systems, i.e., the US Space Shuttle and the Russian Proton and Soyuz rockets. Only the American Skylab space station had a larger diameter of 6.6 m (6.25 m interior diameter) because it was constructed from a converted Saturn V stage. A newer approach to provide for larger volumes is the use of inflatable habitats (cf. TransHab, Bigelow[8]). In addition to an extended interior diameter due to its compact transportation and deployment on-site, those habitats can take advantage

[7]Basically, 1959 requirements [for the Mercury astronauts] were: Less than 40 years of age; less than 5 ft. 11 inches (180 cm] tall; excellent physical condition; bachelor's degree or equivalent in engineering; qualified jet pilot; graduate of test pilot school, and at least 1500 hours of flying time. https://science.ksc.nasa.gov/history/early-astronauts.txt.

[8]TransHab was a concept for an expandable habitat by NASA in the 1990s, intended to replace rigid modules and to expand the inner habitable diameter. In 2000 the private company Bigelow Aerospace has purchased the rights to the patents and has developed the concept further. The BEAM module (Fig. 3.1) has docked to the ISS in 2016. The B330 inflatable habitat is being developed for Lunar and Mars missions.

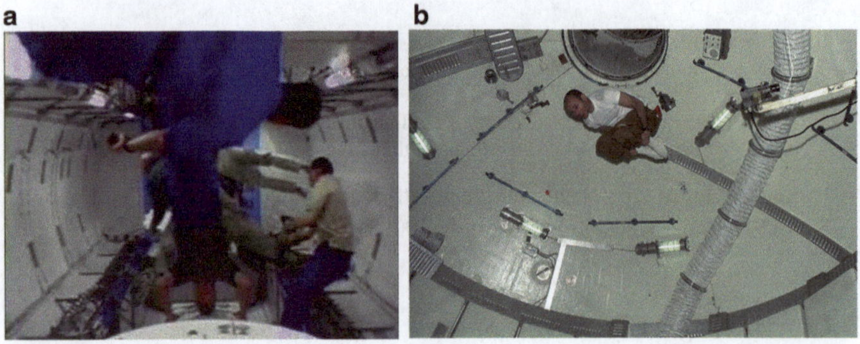

Fig. 7.4 (a) Astronauts enter Kibo module. Astronauts enter and fool around in the newly installed Japanese Kibo module in 2008 (Credit: NASA). (**b**) Early acrobatics onboard Skylab. Astronaut Alan L. Bean, Skylab 3 commander, doing acrobatics in the dome area of the Orbital Workshop (OWS) on the space station cluster in Earth orbit. The dome area is about 6.7 m (22 feet) in diameter and 5.8 m (19 feet) from top to bottom (Credit: NASA, Image ID: sl3–123-2635)

of in-situ-resource utilization on the Moon or Mars (e.g., covering/sintering the inflatable shell using lunar regolith or Martian soil).

When the Japanese module, Kibo, arrived at ISS in 2008, it represented the largest single ISS module and consisted of six major elements (a pressurized module, exposed facility, a pressurized experiment logistics, an exposed experiment logistics module, a remote manipulator system and an inter-orbit communication system). The Kibo module (and all other standard ISS modules that have been transported using the Shuttle) has an inner diameter of 4.2 m without racks. Upon opening, before the storage racks were installed along both walls, the module was immediately filled with eight astronauts, eagerly taking advantage of the potential of 3d, obviously having a lot of fun (Fig. 7.4a). Similar responses could be observed from the very first day, whenever an adequate-in-size-module was in space. Beginning with Skylab (Fig. 7.4b), astronauts have been exploring the pleasures and challenges of **microgravity**.

Skylab's large open volume provided—for the first time—the opportunity to document the neutral body position in space. The neutral body posture (or Zero-G-posture) is when all muscles are in their neutral state and no extra energy is required (see also Fig. 7.2b).

A microgravity environment presents engineering challenges to most of the equipment and technologies (e.g., fluid behavior and flows, heat transfer etc.). It also has severe physiological effects for the human body, including calcium loss, fluid shifts, skeletal changes, muscle mass loss and vestibular alterations (NASA [MSIS] 1995, p. 178). On the positive side, in addition to its recreational opportunity, the lack of gravity provides a 360° aspect that has unique advantages for spatial usage.

All astronauts have pointed out the potential of this unique extraterrestrial opportunity. "*You don't live on a surface, but in a volume*" (Haigneré 2009[9]) and "*Weightlessness is your friend*" (Clervoy 2009). So far, the advantages of microgravity have only been exploited in minor applications where sufficient volume can be arranged. The **vertical arrangements** of sleeping quarters on Skylab (Fig. 7.5a) and Mir space station, as well as the radial arrangement onboard the ISS (Fig. 7.5b) are a good example. When no sleeping quarters were provided, astronauts simply fixed their sleeping bags in a preferred location and bunked down. These locations were chosen largely on availability of unused wall space regardless of up/down conventions taking advantage of the 360° characteristic of microgravity environments. There turned out to be limitations to that approach, however. Astronauts reported noticeable experiences of disorientation when attempts were made to place workstations and other activity areas in orientations that were discrepant to conventional up/down expectations. However, as astronauts transitioned from short duration missions to longer duration, familiarity and repeated exposure appeared to mitigate such experiences. Jerry Linenger commented that "*At a one-month point, I became a spaceman. I was three-dimensional, I was as comfortable flying as I am walking...I saw volume, I didn't see floor—100 percent spaceman*" (CNN [Linenger] 2000).

This opportunistic utilization of space also exemplifies another characteristic of ICE habitat use of volume—the principle of **one space containing multiple spaces**. The sleeping place of astronaut Reinhold Ewald (Fig. 7.5c) not only was located on the wall of a scientific module but it was also occupying space along a transition path as well as serving as a private place.

This use of the same space for multiple purposes reflects the spatial concept elaborated by philosopher Michel Foucault (Foucault 1971) of *Heterotopia* (hetero—other; topos—place), describing spaces that have more layers of meaning or relationships to other places than immediately meets the eye. The demand for multipurpose functionality of all available spaces in extraterrestrial habitats means that heterotopia must be thoughtfully incorporated into living volumes by design rather than haphazard happenstance.

7.3.4 Making the Most of all *Resources*

To begin, we simply must do more with less. The expense of resource provisioning from Earth or in-situ production will demand levels of efficiency and maximum utility rarely encountered in human history. Invoking heterotopia, every 'solution' must address multiple things (Box 7.5). Every element must be multipurpose, flexible for reconfiguration, and amenable to being repurposed upon demand. Solutions cannot take up too much room or power, generate temperature

[9]Haigneré, Jean-Pierre. 2009. Transcript not published. [interv.] Sandra Häuplik-Meusburger. Paris, France, June 2009.

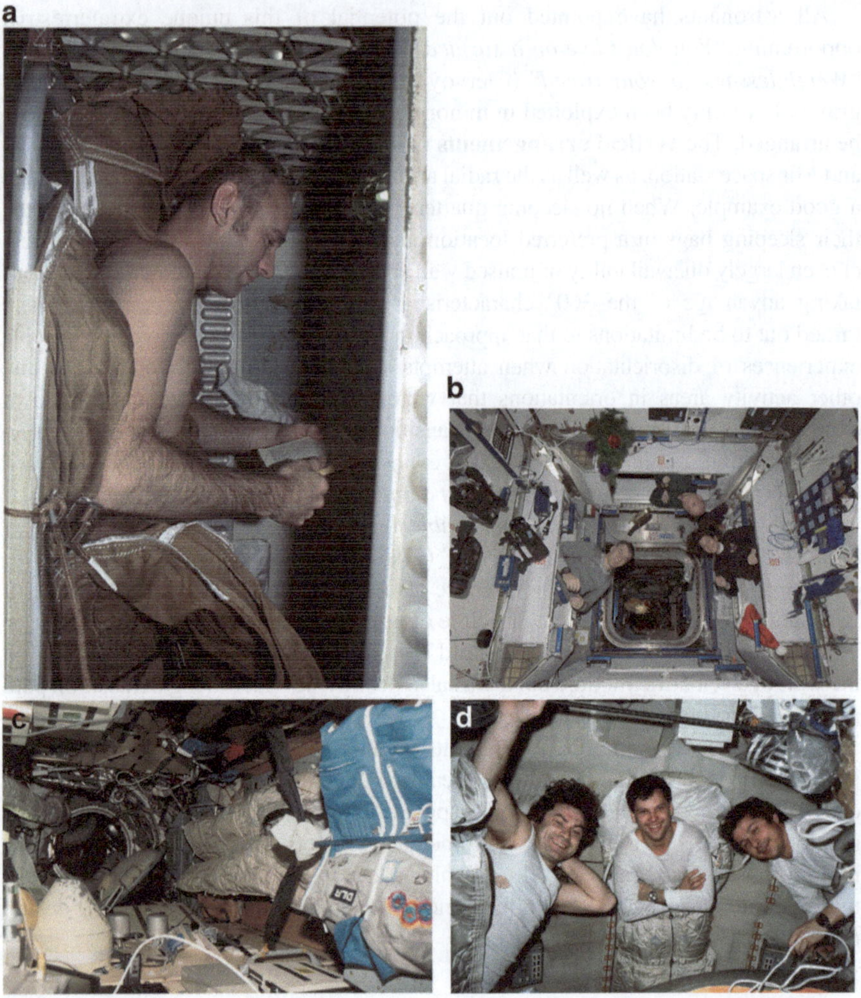

Fig. 7.5 (**a**) Vertical sleep station on Skylab. Astronaut Alan L. Bean, Skylab 3, in his sleep compartment (NASA). (**b**) Radial layout of Sleep station on ISS. Three of the six crew members aboard the ISS peek out of their sleeping quarters on Christmas morning. 2010. (NASA). (**c**) Sleeping along a Transition path. Reinhold Ewalds sleeping place in the Kristall module onboard the space station MIR (Credit: DLR/Reinhold Ewald). (**d**) Sleeping on the Ceiling. Cosmonauts Kovalionok, Prunariu and Savinykh in their sleeping bags on board Salyut 6 (Credit: Dumitru-Dorin Prunariu)

fluctuations, vibrations, or environmental disruptions, require extensive maintenance or attention or interfere with the operation of other critical systems. No wonder we have had only sterile, purely functional operational habitats to date!

From our terrestrial analogues we have had the opportunity to identify and put these principles into practice. For instance, the fact that Antarctica is cut off for 9 months out of the year has demanded a culture of 'making do' with existing

Fig. 7.6 McMurdo station lead baker John Burkart prepares chocolate desserts for the annual Midwinter meal celebration (Credit: Ken Klassy, NSF, 2008)

provisions (Box 7.4). From food to engineering, when supply cannot meet demand, Antarctic personnel have established reputations for inventive substitutions for nearly everything. Terrestrial analogues have helped define needs before we realized they were needs. They have also provided opportunity to try out creative solutions. "*You have to use what you've got in the store—frozen stuff, tinned stuff, and, if you're really desperate, the dried stuff,*" Alan Sherwood, a chef at the Rothera base in Antarctica, told Reuters in 2009 (Kotecki 2018) (Fig. 7.6).

Box 7.4 You have to use what you have got

In 2009, the chef of the British Antarctic Survey's Rothera base, Alan Sherwood, led a team that cooked meals for up to 100 people every day in the Antarctic summers under severe restrictions—the closest food shop is more than 1000 km away across the Southern Ocean in Chile. He had to rely on ingredients shipped in twice a year, in December in mid-summer and in March before winter sets in. That means the food was frozen, tinned or dried. He was brilliantly inventive—a typical menu included pea and ham soup, chicken with pesto, fish in batter, rice, chips. Occasional flights from Chile in summer can bring extra, fresher foods but there are extreme restrictions on weight

> You look out the window in the UK and you've got last night's empty wine bottles and black bags in a dumpster. Here you've got icebergs rolling by," Alan said of his view from the canteen kitchen in Antarctica. "The view changes every day in round-the-clock daylight with huge chunks of ice in the azure bay with mountains in the distance. With supply lines so stretched, you have to use what you've got. Stuart [Mc Dill] and I camped out in a tent one night as part of our training and the sardines we ate came from a tin marked "best before 2004"—five years before our visit. They

(continued)

Box 7.4 (continued)

were fine. The base threw a party for Stuart and me when we left—one glaciologist found what he reckoned was the oldest block of ice floating by the shore—probably more than 1,000 years old—and broke it into drink-sized chunks. He said that old ice has the fewest bubbles in it because they've been squashed out by the weight of snow and ice accumulated above. At that party we drank the base's entire year's supply of tonic in gin and tonics. When we flew back to Chile we had to go to a supermarket to buy a case of tonic—that weight was deemed to be essential supplies, and flown back to the base in Antarctica.

Email conversation with Alister Doyle, who spent 2 weeks at the British Antarctic Survey's Rothera base on the Antarctic peninsula in 2009.

For our off-Earth ICEs, by necessity, efficient use of resources has been an inherent focus from the very beginning. Salyut and Mir space stations were the first stations to incorporate water recycling systems using a system for collecting condensed water from perspiration and respiration out of the cabin atmosphere. Russian scientists also worked on urine recycling technology, but apparently it was never used in space. However, in May 2009, astronauts on-board the International Space Station celebrated a milestone towards self-sustainability by drinking water that was recycled from their urine, sweat and water condensed from exhaled air (NASA Press Release 2009).

Box 7.5 Imaginative conversion of something that doesn't work as planned

Full body cleansing facilities for use in microgravity were developed for the Salyut, Skylab and Mir space stations. During the Salyut (3 –7) missions and on Skylab, a deployable shower system was used. As a request from cosmonauts to have a shower "*which is always assembled and ready for use*" (Bluth and Helppie 1987, pp. I-83), the idea was later implemented for the Mir space station. This shower was the first shower with a rigid cylindrical metal shell. In comparison to the other showers it was easier to clean and provided more privacy. However, this shower concept did not satisfy the cosmonaut's needs either. Eventually, it was converted to a steam room onboard the station and later discarded. Overall, the attempt to transport the shower concept from Earth to space failed. However, in terms of future design developments the 'sauna concept' of the Mir space stations was an imaginative alternative. (Adapted from Häuplik-Meusburger 2011)

The demand for maximal utilization is an inflexible characteristic of any closed loop system. When the unit is located in circumstances in which replenishment and reprovision is costly, difficult and involves extreme delay, system efficiency must strive for 100% even if that is pragmatically impossible. Designing for these high goals must include not only the engineering and technological aspects but the human aspects and their interaction with the living environment. The conversion of the shower system into an easier usable steam bath by the cosmonauts is an example of inventive conversion when resources are limited (Box 7.5). The authors are convinced that many human psychosocial needs can be met with effective habitability design (see especially Sect. 7.7).

7.3.5 *Designing for a Limited Social Environment*

The fact that it will be a considerable span of time before we have communities the size of McMurdo off-Earth means the negative impact of limited group size will be with us for a substantial length of time. It is not an inconsiderable issue. Challenges of living in a microsociety with the outlook of extraterrestrial habitats, have been formulated by Connors, Harrison and Akins (Connors et al. 1985), and include "*intense contact with very few people*" as well as "*acquaintance processes that "are [...] forced, rapid, and relatively unbalanced by alternative relationships*" (p.10).

A metastudy by Bell, Brown and Mitchell (Bell et al. 2019) on teams in long duration space exploration missions conducted in analogues included 72 sources incorporating 253 effect sizes and 1150 data points. Their results suggested that team cohesion and performance relationships in ICE environments may be different enough than those identified in other non-ICE groups (e.g., military teams) that evidence from teams in non-ICE situations could not reliably be used to inform us about the relationships and needs of teams in ICE conditions! These differences centered around five themes:

1. Team members in longer missions generally spend **less social time** together than during shorter missions;
2. Consistent team efficiency over time was typical, whereas decreased team efficiency over time was atypical;
3. By 40% of mission completion or 90 days, all teams reported **at least one conflict**;
4. Commanders' written communication with mission control decreased in length over time; and
5. Team mood dynamics did not consistently support a third-quarter phenomenon[10] (Bell et al. 2019, p. 1).

The implications from the trends have direct impact on design for extraterrestrial habitats. For instance, the propensity to spend less social time together as duration increases (theme '1') means there will need to be the capability to **create more social 'space'** to enhance and limit social distancing over time, while at the same

[10]The **third quarter phenomenon** refers to the decline in performance during the **third quarter** of missions in isolated, confined, and extreme environments, regardless of actual mission duration (Van Wijk 2018).

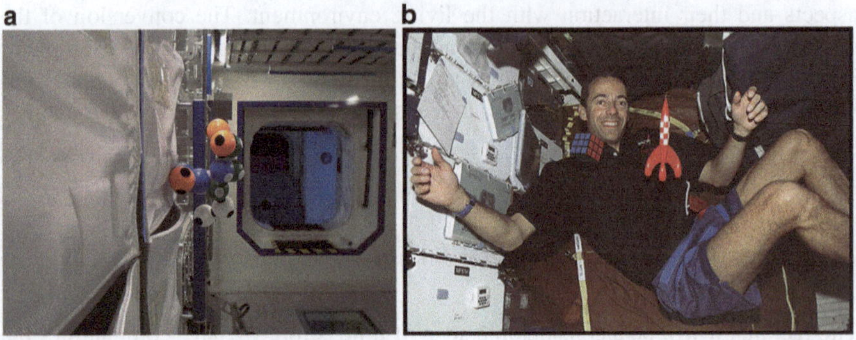

Fig. 7.7 (**a**) *Game structure intermediately fixed to existing Velcro onboard.* As psychological countermeasure in form of a physical provision in enhancement of the spacecraft environment, the game presents a socially interactive and manually and mentally stimulating leisure experience that can contribute to individual relaxation and crew cohesion. (Published in Häuplik-Meusburger et al. 2010a, b). (**b**) Astronaut Clervoy onboard the Shuttle with a Rubik Cube and toy rocket. "*(...) the human body and spirit are uniquely free to float and reflect high above Earth, in weightlessness. We should learn to adapt to this new physical and psychological environment because there resides our far future.*" (Clervoy 2009) (Credit: NASA, Jean-Francois Clervoy)

time **provide private space to meet growing psychological needs** for both over the course of the mission. It could take the form of leisure activities (e.g., games or sports) and include technology to spend time away from the group, e.g., in virtual communities (Fig. 7.7a, b), which also presuppose specific spatial and technical demands. The design of reconfigurable spaces could be used for a number of different social activities in addition to temporarily host individual hobbies that also contribute to group needs (e.g., gardening, robotic exploration, resource acquisition or inventions). Perhaps deployment of inflatable additional modules will be scheduled at critical developmental intervals, for similar reasons as the 'surprise gifts'[11] delivered to the early cosmonauts. Even proposed scheduled shuffling events such as reassignment to different sleeping quarters (aka 'moving day') that give different external views and access to communal resources (laundry, exercise equipment, transition routes) may be employed to instill perceptual 'refreshment' of the environment.

The certainty of increased opportunities for conflict over time (theme '3') demands effective group management. In most terrestrial environments, distancing oneself from others in conflict is a readily available strategy. You leave your work group at the end of the day to return to your own family and friends. There you have the luxury of venting to sympathetic others and seeking social support. In an ICE, your work group is also your entire social group. There is no 'leaving'. It is impossible to avoid interaction. Mission task demands and role demands will require ongoing cooperative interaction. **You can't leave. You can't avoid. You can't**

[11] Surprise gifts have been routinely included for the cosmonauts onboard Progress supply vehicles that serviced Mir space station.

escape. Instances of conflict and the subsequent interpersonal tension must be addressed and effectively resolved. The habitat must be designed such that spaces for small groups are readily available that provide full privacy (shielded from sound and vision). In simulation habitats, all kinds of spaces that provide for 'micro/psycho/hygiene conversations' are used. Oftentimes, it is the airlock, a side module used for storage or any space that can accommodate two people; even EVAs are used to 'get some fresh air'. Creative alternatives have been observed. For instance, intra-group 'conversations' between crewmembers via text or email while in the presence of non-included other crewmembers have provided a mechanism for venting privately while physically publicly present. In addition to psychosocial behavioral skills and approaches, resolution of tensions may be fostered through the provision of activities that require substantial physical exertion, interdependence, and/or cooperative interaction (e.g., a sport or challenge, or games) in which success generates feelings of fellowship and inclusion. Multiple provisions for such activities must be present in the habitat design.

Even the propensity for commander communication to decrease over time (theme '4') has design implications. There is an inherent dichotomy in the commander's responsibility to mission authorities and to their own crewmembers. As distance and mission duration lengthens, priorities and loyalties have a tendency to shift from 'them' to 'us'. However, the need for timely, mission critical information is ongoing. Dependence upon human willingness to convey critical information places terrestrial oversight groups, e.g., mission controls, at risk for receiving inaccurate or incomplete data. Automated data reporting capabilities must be hard-wired into the systems to ensure timely and accurate information. Even with time lags, the ability to query extraterrestrial systems needs to be guaranteed and not dependent on the compliance of the crew. However, such provisions frequently raise feelings of transgressions on personal 'rights' to privacy and autonomy, i.e., the perception that 'Big Brother' is watching! These thorny issues must be addressed and established prior to crew selection and be part of mission inclusion. Reporting should be transparent and viewable by both crew and command and control. There should be a protocol for resolving items of concern raised by either party. Technology that offers greater 'telepresence' will likely improve communication efficacy and conflict resolution with external others. Building in monitoring for the purposes of crew support will walk a very thin line between intrusive oversight and helpful support. The more automated and passive it is, the more out of mind it becomes.

Finally, limited group sizes represent limitations across every dimension of social interaction and interpersonal needs (see also Sect. 7.7). Friendships, intimacy, shared perspectives, interests, opportunities for providing support and meeting emotional and psychological needs are limited to the members in the group. There must be enough attitude and value similarity to provide for group compatibility and cohesion but enough variability to provide for creative interaction and problem-solving. In Living Aloft (Connors et al. 1985), authors proposed that "*... it is essential for crewmembers to share certain general values and desirable for them to exhibit variability in the ways that these values are expressed.*" They noted that larger groups were not as prone to incidences of hostility and conflict as smaller groups because there were greater opportunities for different individuals to provide for psycho-social needs.

In small groups, the limited pool of available individuals represents another source of finite resources. The ability of an individual to provide for such needs means they will become a sought-after resource for others. This may present an unbearable demand upon such crew members. During the EXEMSI chamber study in 1992 (Kass and Schiemann 1996), a crew consisting of three men and one woman successfully completed 60 days of isolation and confinement. Prospectively selected for psychological compatibility, the team exemplified superior performance without many of the intracrew communication difficulties and interpersonal conflicts observed from prior studies. Upon completion, crewmembers were asked whether they would be willing to 'fly' with each of their fellow teammates again. All male crewmembers gave glowing affirmations of each other and for their fellow female crewmember. However, although she enthusiastically endorsed each of her fellow male teammates, the sole female team member noted that all of the other team members frequently sought her out to mediate the usual interpersonal conflicts that arise in any group. As the emotional support person for everyone else, it was an exhausting role (private communications). Subsequent research on gender composition for ICE groups has reaffirmed this experience (Leon and Sandal 2003; Bishop 2004; Kanas et al. 2009; Bishop et al. 2009). Women have been traditionally perceived to be the go-to persons for interpersonal support (known as boundary role persons or liaisons) regardless of whether they are personally desirous of such roles (Bishop 2004; Kanas et al. 2009; Bishop et al. 2009). Thus, gender composition for future teams must include equitable representation to spread the supportive group functions.

On Earth, for the most part, if we don't like a place or person, we can leave that place or person and go elsewhere. In ICEs, especially those future habitats in space, that will not be as easily accomplished (Box 7.6). Therefore, the need for effective countermeasures to be designed, incorporated and implemented into living spaces to deal with conflicts, promote interpersonal accord and provide for psychological and emotional needs is critical. Future space groups will be 'stuck' with the companions they leave with. The physical space has to be designed as such to allow for conflicts.

Box 7.6 One day late!

During a polar winter-over, a physician related a situation whereby a crewmember decided that he could not deal with the enforced isolation and confinement the day AFTER the base closed for the winter (Private Communication, Bishop). Faced with the non-negotiable reality of coping with the isolation and confinement for nine months, the crewmember showed up to lament his misery every Wednesday for an hour with the team doctor. The physician related his struggle to remain empathetic with the crew member which generated feelings of dissonance and guilt in his role as a health care

(continued)

Box 7.6 (continued)
provider. This negative spiral was being enacted with others on the team also struggling to deal with the disgruntled crew member resulting in social ostracization and marginalization. Psychologically, it generated a very distressing and potentially dangerous team dynamic.

7.3.6 Summary

What should be evident by this point is that we must approach habitability for extraterrestrial locations not only from the viewpoint of scarcity but one of *finiteness*. On Earth, the issue of available resources is more often a matter of accessibility, i.e., how much is readily at hand, how hard is it to get more. In extraterrestrial environments, obtaining 'more' may not be possible, much less simply difficult to obtain. On a vessel in-route to Mars, a significant puncture in the ship resulting in loss of atmosphere can be catastrophic regardless of how much safety margin was originally built into capacity. There will simply be no available resources to replace the lost breathable resource. It is not a question of where to go to obtain more. Your destination may literally not have the resource available. This aspect of finite resource availability versus fixed resource demand dictated by extraterrestrial existence was chillingly exemplified in the science fiction story, Cold Equations, by Tom Godwin (1954). Faced with the unreconcilable reality of a finite resource (fuel) and a fixed demand (distance), the freedom to make alternative choices to putting a stowaway out the airlock was zero. A mindset of finiteness is mandated when dealing with extraterrestrial environments because *everything* must be provided for in the closed loop habitats needed to exist off-Earth. Until we locate (and settle) Earth-like exoplanets where resources are, once again, a matter of accessibility, finiteness will rule all things related to surviving and thriving off our birth planet. The inescapable reality of objective, human and spatial finiteness requires intelligent and forward-looking open planning. Again, it is not the question of how much technology you put in, but what people can do with it.

7.4 Camping Versus Residing

> It was a little bit cramped, we had to sleep in small places; it was a little bit like camping and, of course for eight days on board the ISS you can for sure do that. (Frank de Winne, interview, 2009[12])

---

[12]Häuplik-Meusburger S., interview with Frank de Winne in 2009 [transcript not published].

ESA astronaut Frank De Winne's comparison of his experience to 'camping' illustrates the different requirements and expectations people have for long-term versus short-term space missions. The distinction between camping and residing has been referenced multiple times and while it is intuitively understood, it's worth taking a moment to explicitly discuss the differences between the two. Especially, as we have argued, since most existing ICE habitats are only minimally more than (technologically) upgraded camping facilities.

Generally speaking, *camping* is considered an outdoor activity consisting of more than one overnight stay away from home in a temporary shelter, such as a tent or a recreational vehicle. This definition incorporates the presumption of a permanent place of residence elsewhere. The key aspect is that the period of activity is of finite duration and any facilities employed are also used for a transient or temporary stay.

To *reside*, on the other hand, is usually another term for to 'dwell permanently' for a considerable time; to have a settled abode for a time; or to remain for a long time (WikiDiff 2020). Cultures with itinerant and nomadic lifestyles or lack of permanent dwellings are not considered to be camping as transience is not a temporary aspect but a part of their way of life. However, even in such cases, there exist established forms of shelter (e.g., caravans, yurts) that are simply mobile rather than static.

So, the differentiating characteristic between camping and residing is the aspect of permanence in use of a shelter of some kind. People seem to be ok with either of them as long as expectations correspond with the perceived reality. Distress arises when, for example, the reality of the (perceived) social and built environment is not aligned with the person's expectation of dwelling. The following two examples illustrate these principles.

7.4.1 Living in McMurdo Station

> An Antarctic research station is like a remote mining town, but because it's nobody's permanent home, it's everybody's home. Dr. Jenny Baeseman, a polar researcher and the executive director of the Scientific Committee on Antarctic Research (SCAR). (Shields 2017)

In all respects, the American Antarctic base, McMurdo Station, represents an ICE settlement. As the largest community in Antarctica (Ross Island) in number of inhabitants in both summer (~1000 persons) or winter (~200–250), it also has the most extensive facilities with more than 100 buildings (from a small radio shack to three-story structures) spread over 660,000 m^2 (164 acres), its own harbor, two aerial landing strips and a helicopter pad. The station consists of a complete infrastructure housing repair facilities, dormitories, administrative buildings, a fire station, post office, chapel, power plant, water distillation plant, wharf, stores, warehouses, and labs linked by above-ground water, sewer, telephone, and power lines. It has amenities such as ATMs, a golf course and two bars (Pattara 2016; NASA, McMurdo Station 2018). The new master plan for McMurdo proposes to condense the current sprawl into a 28,000 m^2 (300,000 ft^2) campus condensing the dining area, gym, laboratories, living quarters, offices, post office, gym, auditorium, and food prep areas into a collection of six core buildings (Schires 2017).

By all definitions, McMurdo might be considered an almost perfect analogue for future space settlements. Except for one more thing. By treaty, there are no permanent residents in Antarctica and the entire population of Antarctica (~4000 in summer and ~1000 in winter across all bases with another ~1000 onboard research vessels conducting research off-shore) are there temporarily, on seasonal or year-round contracts. Even Chile's Antarctic base which contracts entire families for three-year stays is considered a temporary assignment. Keeping in mind that the lack of a sense of permanence of place reflected in many anecdotal reports by Antarctic personnel generates both fission (divisive) and fusion (cohering) influences, do we design for transient or permanent residents (see Sect. 7.7)? How future space bases will handle the assignment of workers will likely be driven by the economics of transporting workers to and from the bases (e.g., Lunar vs. Mars), the size of the base and the desirability of individuals to permanently relocate. Initial bases will be very small and staffing will certainly be considered transitory. As an analogue for initial smaller extraterrestrial bases, McMurdo is too large and we must look at some of the other Antarctic bases for guidance. However, once those initial 'foothold' bases get established and base growth can sustain a larger population, the characteristics of the population, and hence, the habitation to provide for that population *will* evolve and must be *proactively designed and planned for*. Since it is highly unlikely that any robust extraterrestrial base will be staffed by an entire population of temporary workers, there will be both a certain proportion of the workforce that will rotate through the bases to get experience and as part of corporate advancements, i.e., desire temporary assignments as well as those with an interest in colonizing our extraterrestrial locations. Unlike historical settlement inducements where land was provided for making the choice to open up frontiers, the dependence on technology to provide all sustainable resources (e.g., water, air, food production, power) will impede independent opportunities for colonization efforts that are not part of centralized programs. The specter of the corporate controlled mining town is far more likely than frontier settlement by individuals.

7.4.2 The Early Days of American Astronauts on Mir

Another example of the severe impact that a 'lack of sense of place' can bring is exemplified in the experiences of the first US inhabitants on the Mir Space Station. While the Russian cosmonauts felt very much 'at home' in the Mir Space station, the first American astronauts did not.

The Mir space station was the first large long-term space station. It became operational in 1986 with 14 years of inhabited lifetime in orbit. The station was planned for two people for long durations up to about a year. Fifteen years later up to

6 cosmonauts had stayed on board at the same time for short durations (NASA History 2020).

Just before seven NASA astronauts each joined two member Russian cosmonauts for Mir missions, Valeri Polyakov spent 438 continuous days on Mir.[13] After 10 years of Mir orbiting, Norman Thagard became the first American astronaut Mir crewmember in 1995.[14] While Polyakov "*felt very good during the whole flight*" (Hall 2015) and "*demonstrated that after a long trip to Mars, an astronaut would be in good enough shape to work on that planet*", Thagard spent 115 days in space and never felt well in the physical and social environment on Mir. He lost an extreme degree of weight during his mission and was emotionally expressive about his isolation distress upon the arrival of his retrieval team. Soon after his return he retired from NASA. Similarly, John Blaha wrote of his feelings of isolation during his stay, resulting in severely disrupted sleep and experienced feelings of depression. "*Isolation is a tough thing*," related Mr. Blaha (Associated Press 1997). To counteract his deepening distress, he changed his preoccupation with how much he missed his family to developing shared interests with his fellow Russian cosmonauts and was able to mitigate many of the negative impacts to his psychological and physiological wellbeing. Yet, he, too, ended his career as an astronaut after his return. Jerry Linenger's book *Off the Planet: Surviving Five Perilous Months Aboard The Space Station Mir* (2000) is full of mishaps and complaints. Like his predecessor, John Blaha, astronaut Jerry Linenger initially had problems building relations or even friendship with his crewmates and succeeded after making a conscious effort to set aside national differences.

At the time of the Shuttle-Mir program, the Mir had already endured double its planned 5-year lifetime. The cosmonauts and astronauts had to deal with an extraordinary number of life-threatening system failures and emergencies including the most severe fire ever experienced aboard an orbiting spacecraft, (Linenger 2000, p. 101–110), failures of onboard systems (oxygen generator, carbon dioxide scrubbing, cooling line loop leaks, communication antenna tracking ability, urine collection and processing facility), a near collision with a resupply cargo ship during a manual docking system test, loss of station electrical power, and loss of attitude control resulting in a slow, uncontrolled 'tumble' through space. The deteriorating condition of the station and constant struggle to address the emergent problems reinforced the sense of dependency on the integrity (or lack thereof) of the closed loop system that represented their extraterrestrial 'home'.

During his stay, Jerry Linenger described the older modules of Mir as "*drab as a Moscow winter [. . .] pockmarked with raggedy holes, the result of losing battles with micrometeorite and debris strikes over the years*" (Associated Press 1997). The only time Linenger reported feeling the space station Mir was his home, was when

[13] 1994 January 8–1995 March 22, longest human-operated spaceflight.

[14] During the Shuttle-Mir missions (1994–1998) Mir was also occupied by seven long term NASA astronauts.

they finally successfully redocked, after a 27 min 'fly-around' in the Soyuz to troubleshoot problems and with having difficulty docking.

> Looking into the station I could see a lone ray of light shining through the port window and outlining the dining table. We had left some food out for dinner. It was the only time during my stay in space that Mir looked warm, inviting, and spacious [. . .] Mir no longer seemed just a mechanical object designed to keep us alive, but home (Linenger 2000, p. 95). Jerry Linenger similarly quit the corps of astronauts upon his return.

7.4.3 Summary

> We had decorated the station with balloons and hung up Lenin's portrait. [On the 65th anniversary of the October revolution] (Lebedev 1990, p. 259)

If one feels that a space or place is simply a borrowed space (and owned by the community), there is less care, attention and ownership for maintenance and improvement. We rarely think of keeping a hotel room tidy but will make the effort to do so with our own homes. The inherent problem with temporary residences that entail extended durations is that the lack of sense of place eventually erodes feelings of privacy, security, safety, and responsibility. To the extent that humans can be encouraged to identify a space as 'theirs', both individual and group functioning will be positively impacted. To that end, cosmonauts and astronauts have routinely engaged in activities designed to adopt their environment, just like they do at home, but had to make do with a minimum of available possibilities, i.e., private images and gifts to "decorate" their (temporary) home.

In architecture and spatial planning there are multiple strategies for the process of identifying a space as 'yours'. The definition of *spatial appropriation* goes back to French sociologist Henry Lefebvre (Lefebvre and Busch 2020). In 1968, in his book *Le droit á la ville* (1968) Lefebvre developed the theory *Right to the City*, which includes the 'right to appropriation', in that residents are invited to (re)discover and shape the undefined and standardized urban spaces. For Lefebvre, the interaction of social participants constitutes space, and without them it is only an empty and monotonous container without characteristics. In relation to living in extreme environments, the act of spatial appropriation seems a way to produce a vibrant and truly exciting living space. Appropriation can happen temporarily, interactively, participatory, or situationally. This further emphasizes the importance of a socio-spatial design process, as the way that physical spaces can be designed to enable or hinder people to experience and engage with the environment.

7.5 Looking out and Looking in: What Makes Us Feel Confined?

One of the considerations for fitting the right personality to the expected environmental experience is to determine/foresee what that expected environmental experience will be! The limitation of actual physical living space is, perhaps, the most recognized characteristic of all closed loop habitats. Combined with isolation, it leads to feelings of confinement and associated feelings of claustrophobia, lack of privacy and lack of motivation. Yet confinement is not simply a result of square meters per person (see Sect. 7.3.1).

7.5.1 Crowdedness: Too Much, Too Close!

The psychological experience of *too little (perceived) volume, behavioral constraints associated with a particular volume or too much chaos* leads to feelings of '**crowdedness**', a term used in Environmental Psychology and related to a number of behavioral, perceptual and psychosocial impacts. It is also an individual and a cultural thing. Think of a party where many close friends are gathered in a small apartment, talking, laughing, and enjoying each other's company. Now replace the gathering with the same number of strangers, doing the same activities. Chances are most people will experience the latter as crowded and somewhat uncomfortable. The volume has not changed. The square meter per person has not changed. What has changed is our personal experiential parameters for *needed social space*. We need more with those less familiar. Or take the social distancing requirements in response to the COVID-19 pandemic. We have all become acutely aware of the proximity of others whereas before the requirements to maintain distance, we were largely oblivious! To the extent that our fellow inhabitants are perceived as different from ourselves, our perceptions of crowding will be affected. In terrestrial situations, we would simply go elsewhere to gain the needed social space. But in ICEs, there is likely not any 'elsewhere' to go. This is why we have to plan for it.

The study of the effects that perceived density has on behavioral, communication, and social interaction is called proxemics, and was coined by cultural anthropologist Edward T. Hall in 1963. In his book The Hidden Dimension (1966) he impressively describes the use of space in social relationships with the example of a circus trainer and a lion. "*A lion in a zoo will flee from an approaching man until it meets an insurmountable barrier. If the man continues the approach, he soon penetrates the lion's critical distance, at which point the cornered lion reverses direction and begins slowly to stalk the man*" (Hall 1966, p. 13). He then translated his observations of the critical distance into the well-known Interpersonal Distance diagram (Fig. 7.8). The term personal space has been defined as an invisible boundary surrounding a person's body *"into which intruders may not come"* (cf. Gifford 2002).

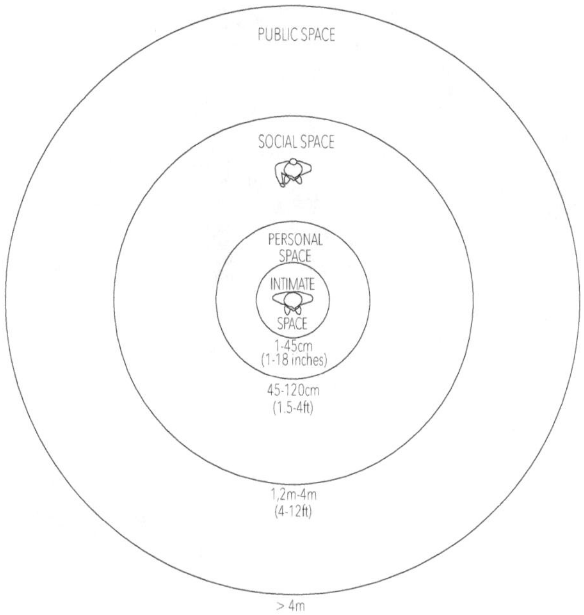

Fig. 7.8 Interpersonal distance diagram. Hall described the relative distances between people with these four zones: intimate space, personal space, social space and public space. Distances are relative and depend upon the individual and society. Drawing by the Authors, based on the theory of Hall

A dramatic recent worldwide example demonstrating the importance of the ability to control one's social interactions and interpersonal presence of others was the social distancing protocols implemented by the COVID-19 pandemic in 2020/21. Social distancing protocols simultaneously limited access to others (isolation) while imposing forced confinement with others (i.e., family). Like teams in ICEs and future space inhabitants, there was the dual stress of isolation and confinement resulting in heightened perceptions of crowding in spaces previously deemed adequate (Box 7.7).

Coupled with our social and individual expectations of personal space, *situational conditions* also experientially define "adequate" space. Spaces that are comfortable in one circumstance suddenly become restrictive under another. Standing in a crowded elevator is completely tolerable until it breaks down and you are stuck between floors. The *inability to leave a space at will* is one of the strongest precipitators of feelings of confinement. Extraterrestrial vehicles and bases will not provide for spontaneous, impromptu exits and strolls around the facilities. Even planned excursions outside require attention to provision of life support and measures to ensure safety and survival (Fig. 7.9). Hence, Antarctic personnel frequently report increases in volunteering for external activities in the months following the closing of the bases for winter. Getting outside becomes more desirable than

Fig. 7.9 Mandatory field-safety and survival class at McMurdo Station is required before anyone can leave for field work or recreation outside (Credit: Elaine Hood, National Science Foundation)

the discomfort or risk of the extreme cold or the demand of the physical activity. In the DOD Unified Facilities Criteria (UFC) for Arctic and sub-Arctic buildings (2004) it was noted that '*[i]n cold weather there are psychological advantages in being able to get away from living and working areas by walking in the covered passageways'* (DOD 2004, 1–1).

Box 7.7 Staying Home During COVID-19

Spring of 2020 will be marked in history as the time when the entirety of the human population partook in an unplanned period of social isolation and confinement due to the COVID-19 pandemic. This extraordinary hiatus in normal human affairs immersed the larger portion of the world's population in a modest experience of prolonged confinement and isolation. What was most extraordinary was not the degree of confinement or isolation (for almost every ICE group experiences these to a greater extent), but the shared commonality of experience as entire segments of the population complied with public health mandated minimized physical social contact to an extent unprecedented in human history. The utilization of electronic communication, entertainment, social media, and virtual networking became primary modalities through which individuals tried to manage the stresses of confinement and isolation from others in much the same way terrestrial and historical orbital crews have done.

(continued)

Box 7.7 (continued)

Various resources were made available from the ICE community to the world at large to help with coping ranging from 'survival tips' to podcasts relating relevant lessons learned from researchers to the community at large (Istrate 2020,[15] Gotian 2020).

To the degree that going 'outside' is not possible, the interior environment will be experienced as more confining if not designed accordingly.

Feelings of crowding are also generated by the *extreme artificiality and visual complexity* of built environments. The interior of existing ICE habitats, including space vehicles or research stations in space, strongly depend upon technology, and are often described as aesthetically neutral and functional, 'cold' in terms of the use of colors and, at the same time, as visually crowded (stimulus overload). This primarily stems from the fact that while initially spartan, as the life cycle of technological environments progress, these minimal interiors typically become totally cluttered; thus exacerbating the existing negative effects of impoverished habitability with chaotic visual crowdedness (Fig. 7.10a, b). Design considerations for habitat use over the life-cycle has been almost never been integrated into the planning process, at least not beyond a pure technical-functional level. During the design process, all space modules have looked efficient and spacious; yet all of them

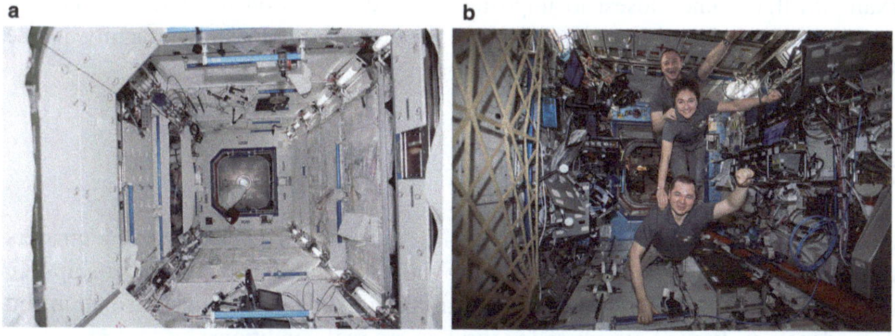

Fig. 7.10 (**a**) Interior view of the Destiny laboratory in 2001. An overall shot of the newly attached Destiny laboratory during the early occupancy by astronauts and cosmonauts from the Expedition One and STS-98 crews, in 2001 (Credit: NASA). (**b**) Interior view of the Destiny laboratory in 2020. The Expedition 62 crew poses for a playful portrait aboard the International Space Station's U.S. Destiny laboratory module. From top to bottom are, NASA Flight Engineers Andrew Morgan and Jessica Meir and Roscosmos Commander Oleg Skripochka (Credit: NASA)

[15]Istrate, I. (Host). July 31, 2020. Sheryl Bishop: The psychological effects of isolation & confinement. Skills for Mars (#18). In *Making Remote Work* series. Organizational Design Community. https://youtu.be/9lgvOGR4IMM.

have turned into cluttered places. The process of this transformation is as much a reflection of changes in mission demands as it is simply the propensity for humans to choose expeditious opportunities for storing items rather than more time consuming and organized systematic efforts. Humans are opportunistic entities and procrastinators as well. Mission planers and engineers have not taken these ubiquitous human propensities into consideration when designing for humans in the machine. The purely engineering approach to designing habitats will, thus, continue to fail to meet 'user' needs (cf. Sect. 7.4.3). In terms of designing a long-term habitat, this approach would be a clear design and planning error leading to inadequate design. Life in a compact environment, where nothing goes in and nothing goes out, needs attention to both the physical needs of the inhabitants as well as the psychosocial needs.

The impact of visual complexity is attentional fatigue. The cost of diverting attention from one stimulus to the next, for instance in a visually cluttered interior, can be substantial (Worden et al. 2003). Walmsley and Lewis (2014, p. 245) found that stimulus overload produced by visual crowdedness can lead to behavioral constraints, such as the inability to handle information and further contribute to stress and anxiety.

7.5.1.1 Terrestrial Examples: Antarctica and Aquarius

With few exceptions, most of the terrestrial analogues are not truly characterized by severe reliance on life support systems and inability to leave the habitats. Two examples that come closest to imposing those demands on inhabitants would be the winter-over bases in Antarctica and the Aquarius underwater facility in the Florida Keys (see also Chap. 5).

7.5.1.1.1 Antarctica

Regardless of size of base, everybody in Antarctica lives in cramped quarters whether it is tents in field stations during the summer, or dormitories or ships. In the winter, only the bases and ships can sustain life, and confinement, isolation and dependency on limited resources for survival are real conditions.

In addition, the emphasis on the transient nature of residency at McMurdo is clearly demonstrated in the briefing each arrival receives:

> Lodging. Rooms are similar to those in college dormitories, and participants are assigned at least one roommate. Roommate requests, including spouses or partners, may not be honored for temporary or transient McMurdo residents. Temporary residents are defined as ASC employees staying fewer than 30 days and grantees staying fewer than 15 days. Those transiting through McMurdo to South Pole or field camps will be assigned to transient lodging. Due to round-the-clock operations, roommates may arrive at any time of the day or night (United States Antarctic Program).

Privacy is also minimal with a variety of shared personal spaces being the norm. A comparison of Antarctic stations indicates dormitory rooming and shared bathroom facilities being very common (Fig. 7.11a, b, also Fig. 5.20a, b).

a b

Fig. 7.11 (**a**) Personnel at the Amundsen-Scott South Pole Station gather in the gymnasium for musical entertainment (Credit: Jeffrey Donenfeld, NSF, 2012). (**b**) Dorm-style single rooms at Amundsen-Scott South Pole station (Credit: Jeffrey Donenfeld, National Science Foundation NSF, 2012)

7.5.1.1.2 Aquarius

For the underwater Aquarius facility in the Florida Keys, the dependency on a functioning closed loop system and limited resources is even more stringent. Situated at 19 m (62 ft) off Key Largo, the depth requires saturation diving protocols and decompression before one can surface. Thus, an emergency exit for any reason represents a complex transit lasting 17 h before reaching safety. Consisting of three compartments, measuring 3 m (~10 ft) in diameter and 14 m (~46 ft) long providing about 37.16 m^2 (400 ft^2) of living and laboratory space for up to seven crewmembers. "*It's about the size of a school bus so it's quite cozy,*" says Mike Heithaus, Dean of Florida International University, which currently owns and operates the Aquarius facility (Loria 2016). The underwater lab is not that different in size from the U.S. Laboratory module on the International Space Station. Astronaut Shannon Walker commanded NASA's NEEMO 15 and observed, "*We're living in small quarters and there are six of us here. So, in that respect, it is very similar to the station.*" She, however, noted a critical difference, "*Of course, the station is actually a lot bigger than the Aquarius habitat, so we have more room on the station and no gravity up there, which gives it a different twist.*" (Pearlman 2011) (see also Sect. 7.3.3) (Fig. 7.12).

7.5.1.2 Orbital Examples: Mir and ISS

Privacy in space stations is, according to astronauts and cosmonauts, less of an issue for short-term missions, but becomes more important over time.

During the Apollo missions, astronauts had no privacy. Sleeping and hygiene activities all in one module, next to each other "*was not a problem*" (Schmitt 2009[16]), as "*most pilots are not used to privacy*". The Salyut (1971–1986)

[16]Schmitt, Harrison. 2009. Transcript not published. [interv.] Sandra Häuplik-Meusburger. Vienna, Austria, 2009.

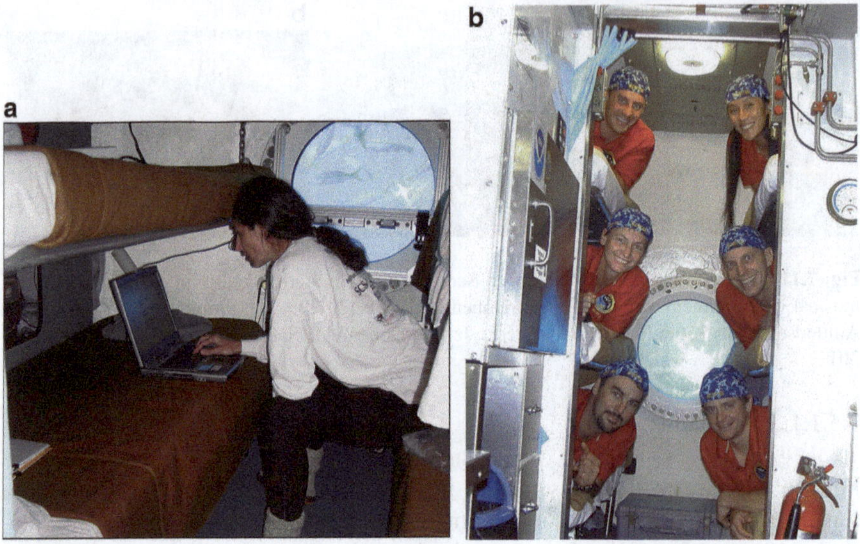

Fig. 7.12 (**a**) Inside Aquarius. Astronaut Nicole P. Stott works at a laptop during her April 3–20 stay inside the Aquarius Underwater Laboratory off the coast of Key Largo, Florida. Stott is a member of the ninth NASA Extreme Environment Mission Operations (NEEMO). (Credit: NASA). (**b**) **Inside Aquarius**. The NEEMO 5 crew members are pictured in the bunkroom aboard the Aquarius research habitat, as their mission winds down. The crew spent 14 days June 16–29, 2003, in the undersea habitat. They are (top, left to right) astronaut Garrett E. Reisman and NASA scientist Emma Hwang; (middle) astronaut Peggy A. Whitson, commander; and Clayton C. Anderson; and (bottom) habitat technicians James Talacek and Ryan Snow, employees of the National Undersea Research Center (NURC) (Credit: NASA)

cosmonauts were not provided private crew quarters and slept on the 'ceiling' of the large-diameter work compartment (Fig. 7.5b). Due to safety reasons, in the early time of cosmonautics, a visual overview of the whole volume was of more importance (cf. Häuplik-Meusburger 2011). But with the goal of future long-term missions, space stations interiors advanced rapidly (Table 7.2). Skylab (1973) was the first space station that provided spatially separated and private crew quarters. On the space station Mir, individual cabins with windows were provided for each of the two cosmonauts (as designed for), so additional cosmonauts and astronauts slept in their sleeping bag somewhere else in the station. Today, the International Space Station provides private crew quarters for six astronauts.

Interestingly, despite the fact that the Shuttle sleep station (see Fig. 7.13) was just made to fit to the body with not much space left, it was perceived positively. "*And if you think, you are in a box, you feel like you are in a box underground ... but if you close your eyes and just feel the weightlessness, you can think, you are in an enormous volume (...) you can imagine you are in a big room. It's ok. It's comfortable*" (Clervoy 2009[17]).

[17]Häuplik-Meusburger S., interview with Jean-Francois Clervoy in 2009 [transcript not published].

Table 7.2 Comparison of privacy levels for sleep and hygiene activities across past and present extraterrestrial habitats; as published in Häuplik-Meusburger 2011, pp. 120–121; 190–191

Station	Privacy of sleep	Privacy of hygiene activities
Apollo	Each crewmember had his couch/hammock; no privateness	'Apollo bags'; No privateness in the module
Salyut	Each crewmember had his sleeping bag; no spatial separation	Toilet and hygiene compartment could be closed off with a rubber curtain
Skylab	Personal crew quarters with privacy curtains; spatially separated from the work area but next to each other	Toilet and hygiene compartment could be closed off; transparent shower in the orbital workshop
Shuttle	Rigid sleeping stations (on selected missions) provide personal space for each crewmember	Toilet compartment can be closed off, but was not soundproof
Mir	Two crewmembers had a personal area that allowed personalization; others chose adjacent modules	Shower had a rigid opaque enclosure; Toilet could be closed off with a door
ISS	Six crew quarters provide privacy for the permanent crew and allow personalization	Hygiene and Toilet compartment can be closed off

Fig. 7.13 Sleeping in a Box. Three STS-107 crewmembers are pictured prior to their sleep shift in bunk beds on the middeck of the Space Shuttle Columbia. From the left are astronauts Laurel B. Clark, mission specialist; Rick D. Husband, mission commander; and Kalpana Chawla, mission specialist (Credit: NASA, STS107-E-05220, 20 January 2003)

Similar observations have been made for hygiene activities. In total, three kinds of showers have been used in space, on Skylab, the Salyut and Mir space stations. The goal was an easy-to-use hygiene facility for full body cleaning. None of the attempts have worked out, and today astronauts take 'sponge-baths'. Only the conversion of the shower on Salyut into a sauna-like facility was used for a longer time (Box 7.4).

Following the 'Apollo-bags',[18] a number of waste management facilities have been developed, used and redesigned. The toilet of the Skylab station was mounted on the wall for economic use of space. There are different reports on the toilet's usability, and using it, is still a delicate topic. However, the process seems to be technically a little challenging at the beginning, but the crew becomes accustomed to the process. In addition, the placement of the toilet within the space station played an important role in the perception of privacy. On Mir, "*for sanitary and privacy reasons, no one was ever enthusiastic about using a toilet two feet from the dinner table*" (Burrough 1999, p. 88).

7.5.2 A Lack of Stimuli: Monotony and Boredom

Paradoxically, another major factor contributing to feelings of confinement is monotony. It may be that the most severe stressors produced by isolation and confinement involve monotony and boredom resulting from low workload, hypo-stimulation, restricted social contacts from family and friends, as well as a never-changing environment. After all, there are no seasons and changing weather and lighting conditions in space, and little in polar regions. Vladimir Gushin, a psychologist at Russia's Institute for Biomedical Problems asserted, "*Confinement on the space station isn't the problem, it's a lack of stimuli*" (Quinn 2009). All environments, no matter how initially novel and stimulating, become familiar. The greater the familiarity, the greater the loss of stimulation, and the greater the increases in the perception of monotony. Anything experienced repeatedly, without variation, will become perceived as boring, i.e., dull, tedious, and uninteresting. Uninteresting environments generate increasing degrees of disassociation, lack of monitoring, and loss of situational awareness, focus, and attention. In closed loop technological environments, boredom produces reductions in vigilance which can easily result in critical indicators and warnings being overlooked. Soviet ground-based experiments even indicated that any change in the environment, even a change for the worse, at least at the beginning is a positive stimulus (Bluth and Helppie 1987).

To frame the continuum between stimulus overload and monotony, Peldszus et al. (2014) used a model of information overload by Klapp (1986) in her study on habitability design interventions as countermeasures to monotony. Klapp proposed a matrix (see Fig. 7.14) with two axes (variety-redundancy and meaning-entropy) and four quadrants: *good redundancy* on the quadrant meaning-redundancy; *good*

[18]The device used was a plastic bag which was taped to the buttocks to capture feces. After defecation, the crewmember was required to seal the bag and knead it in order to mix a liquid bactericide with the contents to provide the desired degree of feces stabilization (Sauer and Jorgensen 1975).

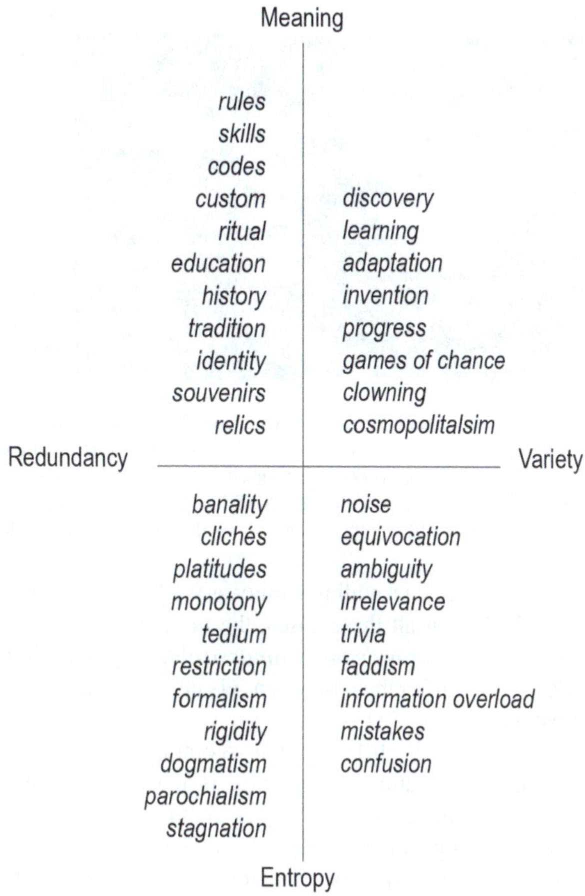

Fig. 7.14 Four sectors of information search. Edgar Klapp developed this model to addressing the significance of boredom as an indicator of overloads of information (Klapp 1986)

variety on the quadrant meaning-variety*; boring redundancy* on the quadrant entropy-redundancy; and *boring variety* on the quadrant entropy-variety. Mæland and Brunståd (2009) proposed that Klapp's interpretative model of how boredom arises may also hold the key to approaching countermeasures to it in an extreme operational context, e.g., deployed military settings. Nominal operations should strive to avoid too much 'boring redundancy' (banality, tedium, monotony, restriction, formalism) and seek opportunities for 'good redundancy' (codes, customs, rituals, traditions, tinkering, continuity, resonance, and 'social placebo' activities such as watching sports) which mitigate boredom as much as 'good variety' derived from discovery, adaptation, invention, games of chance or bantering.

Provisions for good variety and good redundancy can be pursued through features in habitat design. For instance, flexible interior configurations allow for the introduction of novelty and change in an environment characterized by monotony and over-familiarity. Prolonged views to habitat exteriors can provide visual expansion

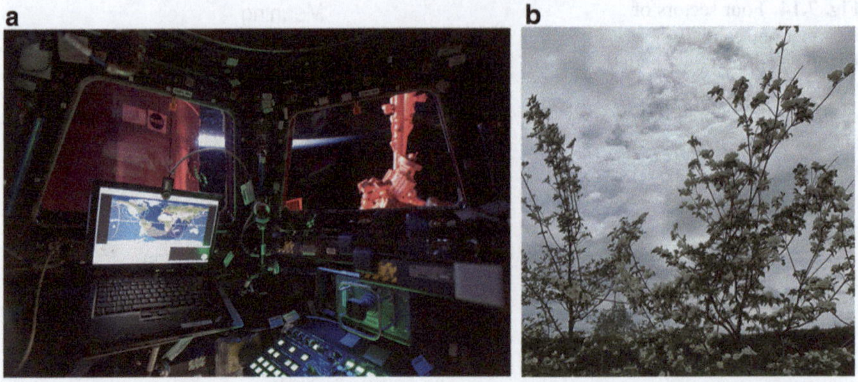

Fig. 7.15 (**a**) Visually crowed environment. This image of the interior view from the International Space Station's Cupola module was taken on Jan. 4, 2015. The Cupola houses one of the space station's two robotic work stations used by astronauts to manipulate the large robotic arm seen through the right window (Credit: NASA). (**b**) Visual complex environment. The twigs of a tree appear geometrically complex, natural and organic (Credit: The Authors)

of interior space, providing variety as well as also countering feelings of containment (Fig. 7.15b). In all these cases, the underlying factor is an increase in *complexity. How is visual crowdedness (undesirable) different from visual complexity (desirable)?* The difference between visually crowded and visually complex can be illustrated by Fig. 7.15a, b.

Complexity is different than 'complicatedness' or the aforementioned visual crowdedness. Natural complexity is made up of abundance, variety, dynamics, growth and change. Complexity is what Klapp (1986) would count as 'good variety', whereas complicatedness is information overload, sensorial or organizationally. Complexity has a positive impact on our sensory system and challenge the intellect—whereas the complicated system, from the technical equipment to complicated procedures and visually cluttered environments, leads to frustration, mental fatigue and ultimately monotony and boredom.

7.5.3 Stretching Boundaries

How do we make confined spaces feel less confining? Although size is not a guarantee for quality, it is easy to imagine that a larger space offers more spatial potential than a smaller space. If used to its full extent, larger spaces are less likely to become boring as quickly as smaller spaces. The addition of more (differing) volume would clearly augment reductions in feelings of confinement as well as provide additional space for activities, safe retreats, increased privacy, improved social cohesion through shared activities and address many other factors.

However, severe limitations on mass and volume have been one of the major drivers for design decisions for space missions. Physical spaciousness requires resources which must also be shipped to base sites, maintained and provisioned with power and life support (heat, water, waste removal) facilities. As exemplified by the example of McMurdo Station discussed in Sect. 7.4.1, not only is the amount of provisions limited by capacity to logistically provide supplies but storage of these provisions places additional burdens upon resources. Hence, in existing ICE habitats, only the minimal supplies and space necessary to store them has been provided. For extraterrestrial orbital facilities, so far, no space station module has had a larger diameter than 6.6 m–6.25 m interior diameter (Skylab space station). To create big space station complexes, such as the Mir space station and currently the International Space Station, a modular architecture approach was applied. It is likely that new future facilities will also utilize a modular approach as a start since it provides for quick expansion with already proven technology as resources become available.

As orbital space vehicle/station capabilities have a feasible limit, deployable and folding structures have been widely used in the space environment for space structures such as solar panels for satellites. Inflatable structures, as the most distinctive deployable bodies for use in architecture, have been occasionally employed and proposed in different forms (cf. Häuplik-Meusburger and Ozdemir 2012). The modular construction approach using prefabricated components assembled onsite with deployable structures is categorized as Class II construction.[19] Advantages are to minimize storage volume and weight during launch and maximize living space and storage at the desired destination (cf. Transhab, Bigelow). For extraterrestrial ground bases, following structures will include built structures with in-situ materials (Class III). Current approaches include the use of regolith for 3D-printed structures, and robotic manufacturing of building materials.

Yet all such expansions will be implemented later rather than sooner and, even once in place, will at some point be similarly experienced as confining. Even a small town can become perceived as overly familiar. Sizing up will not be the complete solution—additional synergistic approaches must be employed to counteract the perception of confinement regardless of the physical volume available.

In addition to physically adding to volume, employment of methodologies to expand the *experience* of existing volume must be incorporated into all designs. For instance, use of configurations that incorporate larger open spaces will foster perceptions of spaciousness as narrow spaces are often perceived to be unpleasant and cramped. Utilizing multiple compartments will provide opportunities to incorporate design variety and segregated activities. Architecturally, use of long views can be incorporated inside the habitat as well. Spaciousness is conveyed through pathing that offers alternatives to travel to the same destination, supporting individual

[19]To describe the different approaches to habitat design, a class terminology is used. Class I construction is based upon pre-integrated modules, like the Apollo Lunar Module. Class III construction uses in-situ resources to build the structures. Class II construction uses pre-fabricated components that are assembled onsite, such as inflatable or deployable structures (cf. Kennedy 2009; Howe et al. 2010).

choice, alternatives to use different routes, increasing occasions to expand social distance and privacy as well as opportunities for greater interpersonal interaction. Supportive use of color and lighting can be employed to reduce feelings of crowding and promote physiological synchronization with sleep/wake cycles. Windows and window analogues have long been recognized as effective architectural elements conveying psychological as well as functional benefits in ICEs and space habitats (Al-Sahhaf 1987). Windows, digital displays or art can be used to counter feelings of monotony and confinement by providing visual depth.

On the Salyut stations, multiple windows were installed for scientific purposes but quickly became favorite places to spend leisure time looking at Earth. For Skylab, the integration of a window in the galley was an explicit habitability requirement (Compton and Benson 1983) (Box 3.1).

Spaciousness can be increased by design and geometry. Sensitive architects and designers can increase the spatial experience of the inhabitants by using textures, surfaces and also kinesthetic senses. A famous and well-perceived historical example is the Japanese design technique of Shakkei. The design principles of ikedori (lit. to 'catch alive') or shakkei (lit. 'borrowed scene') are ancient methods to draw the outer landscape into the garden (Fig. 7.16a, b), that is, to incorporate a distant vista into the composition of a garden or architecture (cf. Neighbour 2001). The aim is to break up boundaries between the garden or buildings and its surrounding landscape to enlarge the perceived physical space. Shakkei gardens achieve this stretching of view with the aid of the vertical and horizontal lines of a building, with hedges, tree trunks or the sky (cf. Häuplik-Meusburger et al. 2011). In interiors, the view is guided with the use of sequential spaces, carefully placed windows or polished surfaces that reflect the outside towards the inside. 'Spacious' houses of famous

a b

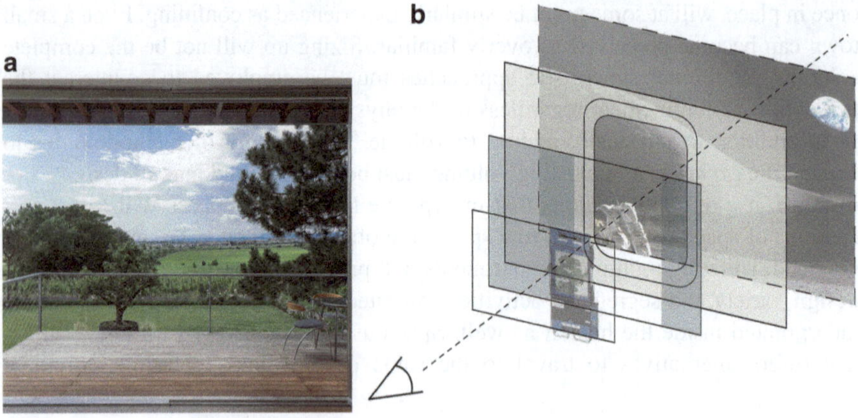

Fig. 7.16 (**a**) Borrowing the view. House and garden designed to 'borrow the view' of the neighboring vineyard, as such the vision is prolonged and the whole plot looks larger and spacious (Credit: space-craft Architektur, Ernst Kainerstorfer). (**b**) Sightline Design. Principle illustrating the incorporation of a distant and natural vista into the composition of a natureless technical interior (as published in Häuplik-Meusburger et al. 2010a, b)

architects, such as Wright, Neutra, Loos and many others reflect this methodology of merging the interior space psychologically with the surroundings.

7.5.4 Summary

> A habitable space is a pattern of effects, not a laundry list of conditions to be satisfied. (pg. 134, Wise 1988a, b)

The experience of confinement is a multiplex phenomena wherein physical, social and psychological factors interact to produce an aversive condition. All closed loop environments will be characterized by features that contribute to feelings of confinement. However, many of these aspects can be mitigated by thoughtful habitability design, activity management and scheduling and provisions for flexibility for inhabitants.

To summarize the factors contributing to confinement:

1. Isolation from family, friends, the familiar social environment and alternative others
2. Feelings of crowdedness
3. Restrictions in going outside
4. Limited and forced social contact with a small group
5. A lack of stimuli leading to monotony and boredom with available environment and social contacts
6. Extreme artificiality and visual complexity of built environments leading to attentional fatigue

Solutions will similarly be multifaceted and serve many purposes. The goal is to introduce meaningful complexity. The addition of more space is highly desirable but not, in of itself, sufficient to mitigate confinement issues. Other approaches must be simultaneously engaged to positively influence the *experience* of the existing volume:

1. Use of configurations that incorporate open spaces that support group functions and social interactions, as well as spaces for small group gatherings;
2. Versatile and adaptable spaces and elements that can be differently themed to provide environmental variety as well as provide for multiple pathing options to enable individual control over social interaction;
3. Use of color and lighting to regulate mood and attention, both individually controlled and ambient;
4. Use of windows (either real or surrogate) as a spatial enhancing tool to provide long views and depth, as well as visual variety;
5. Design of spatial sequences in multi-directions ('up and down') to overview a larger area, provide wider views, visually lengthen the view and enlarge perceived space.

The utility of these approaches and others is maximized by the fact that each will address not only issues of isolation and confinement but serve as partial solutions to many other challenges inherent in ICE habitats.

7.6 Bringing our Own Green: Lack of Natural Elements

Any extraterrestrial habitat will be an artificially constructed environment in the most complete sense of the term. 'Nature' will be radically different from what we think of here on Earth. Most notably will be the lack of flora (plant life) with its abundance of colors, textures, smells, growth and food provision. The significance of discovering even simple-celled extraterrestrial lifeforms notwithstanding, the verdant richness of our birth planet is notably missing for extraterrestrial inhabitation of locations in our solar system. And there are multiple implications for that lack.

One key aspect is the provision of high nutritional, well-balanced and tasty food to maintain health and energy for required activities. Food quality has been repeatedly demonstrated to hold a central importance for psychological and social well-being in all terrestrial and extraterrestrial habitats. Its importance is a ubiquitous commonality across time and human cultures (see Sect. 7.7.3). For that reason, food systems in development for extraterrestrial habitats have been given high priority and advanced a great deal over the last 50 years. It has been shown that food can be grown in microgravity. Small greenhouses are regularly used on-board the International Space Station to spice up the food and for research. It is recognized and understood that greenhouses will need to be fundamentally different the further away we travel. They will have to be (more) autonomous and much larger. Crop hardiness and efficient growth systems will need to be robust and adaptable. Contributions to replenishment of life support systems through the removal of CO_2 and production of oxygen has also been recognized as a beneficial feature of the inclusion of plants into the closed loop system required for extraterrestrial habitats.

Aside from nutritional and life support system applications, there may be additional compelling reasons to ensure the inclusion and integration of green spaces into extraterrestrial habitats. Abiding by the demand to provide for multiple uses and applications, the inclusion of plants into extraterrestrial habitat design may also provide critical noninvasive, passive countermeasures for the deleterious effects of confinement and monotony through sensory and spatial enhancement of the otherwise technical and monotonous space environment (Häuplik-Meusburger 2014). *It appears that not only is greenery restorative but we have been evolutionarily primed for optimal cognitive functioning in environments of particular botanical complexity*.

One key approach to instilling surprise, harmony and complexity could be in appropriating natural design elements for interior habitat environments according to bionomic (the study of the relationship of organisms to their environment) principles. Outside the space exploration field, there is ample thought on the inclusion of natural properties in building and interior design as a strategy for psychological and sensory integration (Kellert et al. 2008). The complex properties of nature have been shown to be psychologically restorative in a myriad of studies due to its infinite variability, evolutionary association with beneficial resources, associations with relaxation and stress reduction activities. Many studies have shown that "natural contents and, in particular, landscape configurations" have positive effects on human functioning, representing sources of desirable visual complexity rather than undesirable visual crowdedness (Joye 2007). The theory of 'Biophilia' or the urge to affiliate with other forms of life, applied to interior environmental qualities argues that this drive is as old as the human species' relationship to the natural environment that nurtured our evolution.

Persistent astronaut reports underscore the importance of plants and living beings during their missions as astronauts have persistently created opportunities to interact with natural stimuli in response to the restriction and artificiality of the habitat (cf. Häuplik-Meusburger 2014; Häuplik-Meusburger et al. 2010a, b). During the Salyut missions (1971–82) cosmonauts experimented with plants and "designed" their own little greenhouse. The benefits were so evident that the Soviets even designed a device "*for the sole purpose of ornamental plant culture to provide psychological comfort to the cosmonauts in the station*" (Porterfield et al. 2003). They also frequently sent up plants and fresh vegetables in the resupply ships. The greenhouse at Amundsen Scott South Pole Station now has a sign-up sheet so that crew members can 'reserve' private time relaxing in the presence of greenery.

In the absence of real plants, astronauts and inhabitants of ICEs, including simulation environments, have put up images of landscapes and natural environments. The notion of 'aesthetics as a function' later proposed by Clearwater and Coss (1991) echoed the space architect Galina Balashova's earlier work on inclusion of natural scenes in Soviet long duration missions (Zimmerman 2003). Bionomics would argue that these substitutions may not be viable substitutions for the entirety of conditions that they seek to mimic. So what is it about plants that is important?

7.6.1 Fractal Geometry and Bionomic Design: Unraveling the Mystery of Greenery!

We are clearly drawn to plants and find innate benefits from their presence. The question is why? A growing body of theory and convergent research from evolutionary psychology and externalized cognition has identified connections between sensed structural patterns of the physical environment and the work performance and quality of life experienced by its occupants. Roger Ulrich's (1983) psychoevolutionary framework proposed that affective responses towards environmental features are rapid, automatic and unconscious adaptive perceptual processes rooted in human evolutionary history in which different environments are immediately liked or disliked in order to facilitate responses that contribute to wellbeing and survival. Studies demonstrating these rapid, immediate and automatic affective responses to various environments (Hietanen and Korpela 2004; Korpela et al. 2002) provide supporting evidence that these emotional reactions saved precious time and energy spent learning which environments were beneficial or harmful (Kaplan 1987, 1988; Ulrich 1983).

In addition to benefits in neural efficiency, natural settings have been demonstrated to produce restorative or rest directed attention (Hartig et al. 2003; Kaplan and Kaplan 1989). Ulrich's psychoevolutionary framework proposed that the restoration from natural environments is a product of being grounded in stress *prevention* that came from the beneficial associations with perceptually benign, nonthreatening and supportive environments. His demonstration of accelerated healing in surgery patients exposed to a small tree group compared to a brick wall provided empirical evidence for the hard-wired nature of the response (Ulrich 1983). Subsequent research has linked responses to natural environments to the subcortical areas of the brain, especially the amygdala, which are involved in modulating stress-related hormones (Parsons 1991[20]; Joye 2007) leading Parsons (1991)[21] to subsequently argue that unrecognized negative stress reactions to post-modern built environments could be responsible for important negative health effects. Joye (2007) proposed including elements of ancestral habitats in the built environment could counter potentially deleterious effects by producing more relaxed physiological and psychological states.

So, there we had the answer, right? Well, as with all 'answers', there is more to the story. The overwhelming evidence that natural environments are, in general, preferred to built environments was slowly followed by intriguing evidence that some natural environments are *more* preferred (and *more beneficial*) than others (Cummins and Cummins 1999; Van den Berg 2004). The fact that benefits have been detectable in very different natural environments while at the same time inconsistently experienced in similar environments prompted some researchers (Katcher and Wilkins 1993), to suggest a "*search for general characteristics of the patterns in nature that produce relaxation*". The resulting exploration revealed a preference for environments with savannah features, a mixed woodland-grassland ecosystem characterized by trees sufficiently widely spaced so that the canopy does not close allowing sufficient light to reach the ground to support an unbroken plain consisting primarily of grasses (Balling and Falk 1982; Synek and Grammer 1998; Orians and Heerwagen 1992; Sommer and Summit 1995). Habitat theory proposed that the reason savannahs were effective was because those low to intermediately complex environments where humankind first evolved displayed an ideal mix of structural landscape features and natural contents (Wilson 1993; Van den Berg 2004).

If we are evolved to process certain natural environments more efficiently, what implications does that have for designing those places where we live? Experimental

[20]Parsons, R. (1991). *Recovery from stress during exposure to videotaped outdoor environments.* Unpublished doctoral dissertation, University of Arizona, Tucson.

[21]Ibid.

evidence for the linkage between cognitive processing and types of environmental patterns involved in these processes suggested new ways that interior design may directly enhance habitability and the performance of 'Knowledge Work' in otherwise impoverished habitats such as those found in most extreme environments (Wise and Taylor 2002) (Box 7.8). This was further fueled by Purcell, Peron, and Berto (Purcell et al. 2001) who proposed an intriguing hypothesis that the beneficial effects of natural scenes might be due to their *fractal* characteristics which were lacking in the Euclidean geometry of traditionally built environments. Fractals are never ending patterns, a curve or geometric figure, that are self-similar across different scales, each part of which has the same statistical character as the whole. Fractal patterns appeared to evoke associations of naturalness in humans and improve the ability to perceptually differentiate between highly similar patterns (Geake 1992).

Investigations into fractal geometry have demonstrated that fractals can create feelings of spaciousness by increasing density and optimizing space without expanding the topological dimension (Hernandez et al. 2010, p. 43). Intriguing as this was as an explanation for why plants improve livability of a habitat, subsequent research strongly demonstrated that preferences for natural settings was *predictable by the underlying geometric, fractal properties* which appear to be responsible for the associated restorative effects (Hagerhall et al. 2004). And, in fact, empirical evidence from computer generated images, human-generated drip patterns (aka Jackson Pollack), natural sources (images and actual presence of flora) (Spehar et al. 2003; Abraham et al. 2003; Hagerhall et al. 2004; Taylor et al. 2005) and even some built environments (Scopelliti et al. 2018) has subsequently indicated that the most effective linkages appear between cognitive and emotional processes and settings characterized by fractal structuring in the 1.3–1.7 dimensional range (D).

Box 7.8 Solving a 16-Year-Old Mystery

A NASA study by Wise and Rosenbeg in 1986 tested the premise that exposure to natural images were beneficial by continuously exposing study participants to one of three different visual scenes (a savannah landscape, which were scattered renditions of acacia trees, dark green on a light background; a nature scene which was a dense wooded mountain landscape, and a black and white pattern of scattered geometric blocks of lines (the control condition)) while performing three mentally stressful tasks (arithmetic, logical problem solving and creative thinking) to induce physiological stress, interspersed with one minute recovery periods creating high and low stress periods. While both natural scenes were beneficial compared to the geometric blocks, the greatest effects (faster stress recovery as well as lower stress reactivity) were found for the black and white savannah image (44%) instead of the expected natural forest scene (3%). In 1986, this result was unexplainable. Over the next 16 years, the evidence for the superiority of the savannah scenes as well as the experimental evidence for evolutionary cognitive preference for

(continued)

Box 7.8 (continued)
fractals prompted Wise and Taylor in 2002 to reexamine those puzzling earlier results by exploring the relationship between fractal geometry and stress reduction. The reanalysis of their 1986 findings included the calculation of D for each scene. The pattern of squares was found to be completely non-fractal, whilst the savannah reproduction and the forest photograph were both found to be fractal with D values of 1.4 (falling within the preferred midrange) and 1.6 (falling outside the mid-range) respectively.

Taylor et al. (2005) noted that the fact that mid-range *D* fractals appear to affect the physiological condition of the observer producing the reductions in stress is exactly as Ulrich's psychoevolutionary framework predicted. He argued that it may be the *fractal dimensions of a pattern, rather than its naturalness*, that is the determining factor for its impact on visual perception and stress-reduction. If so, then incorporating interior design elements characterized by the mid-range fractal properties could produce effective countermeasures to stress and promote well-being, especially in environments where nature's fractals were missing or limited (Taylor 2001, 2003). Thus, a biologically based **'biofractal hypothesis'** of beneficial effects offers both an integrative explanation and a design strategy of significant power and flexibility suggesting that *habitat interiors might be expressly designed to assist innate natural emotional, cognitive and perceptual processes associated with flexible, creative thought, stress reduction and personal emotional management.* Therefore, incorporating these elements into habitat design will provide passive, autonomous cognitive and psychological supportive countermeasures.

7.6.2 Surrogate Views: The Potential of Individual Plant Systems

The time-based qualities of plants (i.e., change through growth) and their distinct color range make them a sensorially different addition to the environment inside the spacecraft. In addition to providing these benefits, plant growth facilities (i.e., green houses) could be used, to an extent, as a substitute for windows, or to make up for a lack of visual information obtained from windows. The incorporation of windows has long been recognized as desirable. Yet we have not had to deal with external views that may lack depth cues or provide changing vistas as will be the case with long duration interplanetary transits which look out into a star field. In such cases, the substitution of virtual 'external' displays may be far preferable. For those locales where an external view looks out onto something, whether actual physical windows or digital 'ports' displaying real-time actuality, the benefit of looking at the outer world argues for the inclusion of the wherewithal to satisfy this basic human need to track the changes in the world around us.

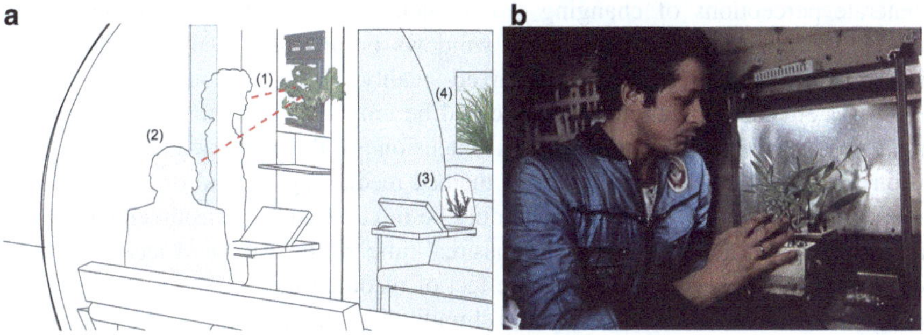

Fig. 7.17 (**a**) Integrated greenhouse design. As a 'window to something living', an integrated greenhouse can provide this surrogate view in all mission scenarios where the inclusion of windows or the view outside is limited: (1) Physical real-time interaction with plants, (2) visual integration into interior, (3) individual plant systems, and (4) virtual window integration (Credit: space-craft Architektur). (**b**) Expedition 5 Cosmonaut Victor Savinykh with 'his' plants on board Salyut 6, 1977 (Credit: TASS, courtesy of Victor Savinykh)

As plants psychologically represent the environment outside of our habitats, views of greenspaces within a habitat could serve as a viable alternative. The relationship to, and experience of, the outside, either through direct egress or indirect observation can address myriad needs in a confined environment. By utilizing plant displays and the virtual displays of natural scenes or materials, the benefits of viewing nature and the experience of 'looking outside' can be provided through 'surrogate views' to achieve restorative effects (Heerwagen and Orians 1986; Levi and Kocher 1999) (Fig. 7.17a).

In more complex missions with a greater number of crew members and a larger habitat, or with sortie missions in semiautonomous vehicles away from a base, the development of small modular growth units could be of advantage. In addition to a large greenhouse facility, small plants could be extracted at liberty and transported in small portable growth units. They can function as 'trabant' (German: satellite or companion) greenhouses for easy access in the galley, act as decor for personal quarters and common spaces or provide live food supplement in pressurized rovers (cf. Häuplik-Meusburger et al. 2010a, b).

In addition to surrogate views vis a vis windows, in multi-module bases or those with distinct sectors, architecturally, use of long views can be incorporated inside the habitat as well. Spaciousness can be conveyed through pathing that offers alternatives to travel to the same destination, supporting individual choice, alternatives to use different routes, increasing occasions to expand social distance and privacy as well as opportunities for greater interpersonal interaction. The central core walls in each module could be designed to represent environmental themes (e.g., evergreen forest, cityscape, prairie, rainforest, mountains, seascapes) incorporating the optimal range of 'biofractal' elements found to facilitate human performance, reduce stress and improve wellbeing. Each module would have a separate theme that is restricted to the confines of each particular structure/sector. The different themes would generate perceptions of changing environments as one travels from module to module. Utilizing digital displays in windows or on entire walls in addition to immersive virtual reality (VR), mixed reality (MR) or augmented reality (AR) systems, such design elements could be employed with minimal impact on power or structural components. An extension of the approach is a proposed Biotrope system comprised of digital, themed media (e.g., sound of the rainforest, waves crashing on the shore) which could be linked to real environments on Earth and updated periodically or in continuous real-time transmission and accessible from each module would avoid the 'loop' effect of recorded tracks. This would provide for greater interior-exterior variety and stimuli within the confines of the station and extend multiple fields of view complete with updatable and flexible presentation. It is conceivable that such mixed reality media could also incorporate specific events or significant places back on Earth as well.

7.6.3 Summary

> The first is the openness in between inside and outside with the possibility of having a view. The second is the ability to reproduce sensorial elements of our terrestrial conditions. And the third one, places for privacy and places for common moments to share.—Claudie Haignere, ESA Astronaut (in response to Chap. 6 survey question asking for three most important factors for a space habitat, 2020).

As discussed earlier, confined environments suffer from a limited range of sensual variety and stimulation. Lack of visual variety and over-familiarity has been clearly associated with decrements in concentration, performance, mood, increases in boredom, restlessness and risky behaviors designed to fill the gap. Given that introducing novelty from external sources is not going to be a readily accessible resource, the challenge becomes how to generate new stimuli from the same environment to which you are currently limited. The central 'bionomic-design hypothesis' proposes that the general spatial and (particularly) fractal structures of benign ancestral human environments act as a template for perceptual and cognitive processes which share control of our perception, memory, and emotional management with the physical environment. Just as basic 'ergonomic design' directly supports physical functional performance, there appear to be pattern, color, texture and spatial characteristics of habitat interiors that facilitate more efficient cognitive functioning and performance as well as promoting stress reduction. The standard of design practice for operational environments has been towards minimalistic, patternless interiors where 'form follows function'. The argument is that such environments are not those in which we have evolved to perform at our mental best. The goal is to introduce meaningful complexity through the use of adaptable spaces (flexible partitions), lighting (diurnal/seasonal dynamics), geometry (fractal patterns, reduced clutter), and providing wide views, i.e., to recreate those elements that compose humankind's 'natural habitat' as proposed by bionomics.

One approach is to take advantage of the increasing realism in digital displays. Visual variety and complexity as well as restorative biofractal stimuli can be digitally incorporated into wall displays or combined with other senses to provide multisensory stimuli. For instance, as part of a multinational project by the British Interplanetary Society (2006) to design a Mars polar base, Project Boreas, in order to provide for environmental variety, addressing various medical needs and counteracting the negative effects of confinement, a Multi-Environmental Chamber (MEC) was proposed that combined visual, auditory, olfactory and tactile sensory stimuli (Bishop 2006). Composed of a small chamber that could accommodate 2–3 crewmembers, it was to be completely programmable for various environments, e.g., a day at the beach would include heat, the smells of saltwater, humidity, UV lamps to simulate sunshine and provide Vitamin D, and sounds of the seashore. By providing a programmable interior, individuals would be able to choose stimuli that meet their personal emotional and psychological needs without imposing their choice upon the entire group. Each programmable environment could be linked to appropriate medical countermeasures (e.g., UV lamps to counteract loss of vitamin D, jaundice, skin conditions; full spectrum lighting to counteract seasonal affective disorder) as well. The capacity to accommodate several people at once would also provide a place for social gathering and stimulate group interaction.

An early example of a programmable interior for the space station is illustrated in Fig. 7.18a, b. Kas Oosterhuis, Emeritus Professor TU Delft and Director ONL Lénárd see all buildings as in principle dynamic structures, consisting of a large set of thousands of programmable components (as published in Oosterhuis 2012).

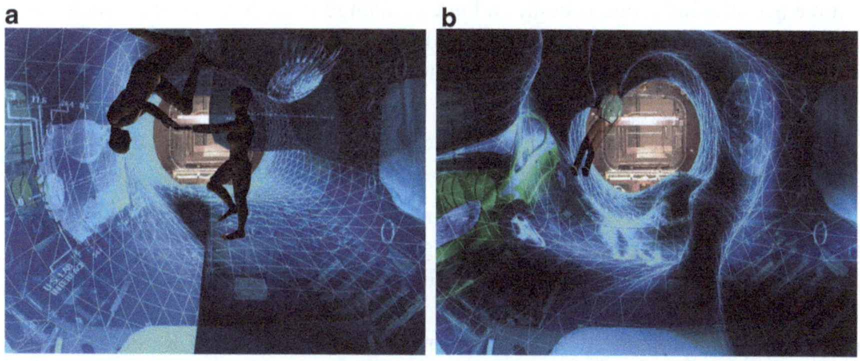

Fig. 7.18 (**a**) and (**b**) Programmable interior for the space station. (**a**) meeting mode and (**b**) home mode. The design vision is to regard buildings as instrumental bodies, which can change their shape and content in real time. Bodies can be addressed, and all constituting components which make up the entire body can be addressed individually (Credit: Oosterhuis, Lénàrd 2000)

7.7 Space Roommates: Social Logic of Space

> That introduces, I think, another element of psychological pressure and stress. The people who are alone by themselves and don't have a friend from their own culture, I think, are going to even have a little more special kind of problem and we need to pay attention to it. (Astronaut John Blaha, Associated Press 1997).

> It's tough not getting any alone time for many weeks at a time, I've always been based on ships, where the work hours are long and the sleeping quarters are close—often four in a very small bunk room. Being alone requires (a) having the time, and (b) having a place to go. Because of safety, you can't always roam the decks of the ship alone, so mostly you are in company. Dr. Nerida Wilson, invertebrate marine molecular biologist, Western Australian Museum (Shields 2017).

Living with others is not easy in the best of circumstances but remove the opportunity to get away when the inevitable irritations flare, shrink the shared space to something slightly larger than a large apartment, and limit the opportunity to interact with anyone else for a protracted period of time (years) and suddenly there simply isn't enough room! Then there are the myriad nuances of social existence to deal with involving cultural norms and practices and individual differences.

7.7.1 When the Need for Privacy Becomes a Territorial Issue

Take privacy for instance. As the opening quote so graphically exemplifies, the simple act of being alone becomes a huge challenge in an ICE habitat. When going outside is severely restricted, the availability of solitude comes at a premium. And with the loss of the capability to be 'separate' comes the loss of privacy.

> "I won't go into all the detail," Mr. Blaha said, "but the bottom line is, awful nice when you're in isolation to have a little bit of private space, even if it's, oh, 7 feet[2,1m] by 3 feet [0,9m] by 3 feet." (Associated Press 1997)

In surprising ways, privacy losses are magnified in confined environments. For instance, while individual cubicle sleeping spaces provide visual privacy, rarely do they provide audio or olfactory privacy. Hence, one's crewmates are forced to share in activities not intended to be shared, e.g., snoring, talking in one's sleep, flatulence, conversations with others. Crewmembers at the Mars Desert Research Station (MDRS) in Utah, routinely joked about the extreme permeability of the thin partitions that separated bedrooms that even permitted music being played with earbuds to be heard by others nearby. The crews frequently resorted to email 'conversations' between individuals physically present while they worked to preserve some degree of privacy in communications. It was not so much that they were sharing information they wanted to keep from other as it was simply conversation that was meant for a specific other and not the group. The experienced loss of being able to control who did and did not overhear commentary, bodily functions or actions became intrusive and prompted efforts to regain such control. One MDRS Commander (see also

Chap. 6) commented that he loved having the commander bunkroom because it had only one shared wall and a porthole which gave him a greater sense of being able to separate from group.

In a 2003 OPP housing report for McMurdo Station, one of the highest housing priorities expressed was a 40% increase in single-occupancy rooms, highlighting the importance of privacy for everyone working at the station. In 2015, the McMurdo Station Master Plan 2.1 (National Science Foundation 2015) proposed scalable available single and double dormitories to adjust for fluctuations in station population between winter and summer demand by reuse of existing structures or replacement if those structures were not usable.

The persistent demand for separate spaces meets needs for solitude and privacy, allow individuals to regulate social interaction and provide for individual control over the amount of contact and shared personal information with others. It provides for individual personalization and individual differences in activities while moderating feelings of crowding and confinement. Privacy is often related to a specific space and to having control over it. When astronaut Jean Francois Clervoy was asked to describe what a home in space would be like, He slammed his hand on the table and said "*This is **MY** bed, this is **MY** kitchen, this is **MY** cupboard, with **MY** stuff in it. **Don't mess** with it.*" (Häuplik-Meusburger 2010, p. 11).

Yet there also needs to be spaces that require and facilitate group interaction. One of the inherent dangers in isolated groups is the general drift of certain individuals into complete isolation from their crewmates. Spaces that enforce human contact and interaction serve to preserve bonds between crewmembers as well as close contact needed for accurate group monitoring and intervention when needed. To achieve these dual needs within the very limited space available in confined habitats, we need flexible, definable and re-definable interior environments that provide for group as well as individual activity. Flexible sleeping stations can be used to personalize the mission crews "home" and thus reflect the community's identity and preferences. Flexible partitions can further enhance usability, to for example, temporarily close off work and sleep areas. For these reasons and many more, separation of private functions from public is critical. Provisions for meeting spaces, leisure and exercise spaces, sound-proofing and securing of spaces for hygiene activities are essential.

7.7.2 *Zoning out Social Conflicts: When Little Things Become Big Things*

In ICEs, it is often the mundane that plays the largest role in interpersonal difficulties. Little things become big things in isolated, confined environments. Bell, Brown and Mitchell (Bell et al. 2019) conducted an extensive systematic review of analogue research and found that all crews studied (from 72 sources) had reported at least one conflict by 40% of mission completion or 90 days.

Even mundane matters such as adherence to hygiene standards have been shown to be important for health and self-esteem as well as enhanced morale and standards of personal decorum and respect for others. A 1993 survey of American astronauts (Santy et al. 1993) found matters of payload/experiment resources, housekeeping tasks and personal hygiene issues produced the highest rates of miscommunication, misunderstandings and interpersonal conflicts in the moderate to high mission impact category.

Then there was the 10-minute bloody brawl between male crewmembers and subsequent attempt to force a New Year's Eve kiss on the only woman crewmember during the 110-day isolation study (SFINCSS) in 1999. In this case, differences in cultural attitudes regarding gender interactions disrupted and interrupted the long duration simulation with crews being physically separated from each other and one participant leaving the project and highlights challenges that will be faced by international crews (Sandal 2004; Mann 2017).

The infamous 1973 Skylab 'mutiny' where the astronauts went on strike in response to overwhelming work schedules was largely directed at Mission Control. Similarly, cosmonaut Valentin Lebedev reported that 30 percent of the 211 days he spent on Mir in 1983 involved crew conflict and "*that misunderstandings among crew members and misunderstandings between crew and Mission Control were the most difficult problems experienced during prolonged missions in space*" (as reported in Mann 2017).

Much of the tension between crewmembers observed in ICEs has been expressed through social (sometimes interpersonal) withdrawal,[22] increased territorial behavior and the use of scapegoats or external others (e.g., Mission Control) (Kanas 1971, Kanas et al. 2009; Sandal 2002). This change in behavior has spatial consequences and is often leading to increased need for privacy with increasing time. Designing for the inevitability of interpersonal conflict poses dual challenges. Preventive measures would be those that minimized areas of potential conflict and irritations, e.g., work stations with adequate facilities, private areas for social distancing, automation for burdensome station-keeping duties, clearly defined personal space and the ability to control that space, restorative areas for relaxation. Mitigation measures would provide for effective means of countering rising tensions—engaging exercise, sports, or leisure activities, opportunities for positive social engagement, provisions that facilitated group fusion instead of group fission. On the habitability side, many features could address both needs. Retreat areas could be partnered with food production tasks in the greenhouse. Pathing through a base should provide multiple routes to reach destinations allowing inhabitants some control over engaging others. Specific areas for various activities like exercise or noisy activities should be separate from quiet areas to provide restoration. The ability to modify social spaces for different needs and events encourages group activities and positive interactions. We will be at both our best and worst and our living space must accommodate those extremes.

[22]In this book we do not refer to social withdrawal in respect to voluntary recreation or exploring one's identity, such as retreats in monasteries, dark room retreats or similar, nor in the sense of the sociocultural mental health phenomena, also called 'Hikikomori' (Neuroscience 2020).

7.7.3 The Path to a Person's Heart Is Through Their Stomach, Fine Manners Help

The statement '*the path to a person's heart is through their stomach*' is an old saying shared by many cultures, implying that food and love go hand-in-hand. Cooking and eating are high-esteemed traditional and enjoyable activities. The art of cooking and the style of having dinner together vary widely throughout the globe and often is the 'glue' within societies. The place where you cook and eat is often the social 'warm' center of the house.

The importance of food to individual and group psychological wellbeing for long duration space crews cannot be overstated. Meals are central events and food and eating are major sources of stimulation and variety for all ICE groups, including space. Meals promote orientation to group welfare, provide opportunities for face-to-face social exchange and associate satiation of physiological needs with positive group interaction. Provisions should be in place as much as possible for individual selection, snacking, and protocols for the sharing of food preparation tasks, i.e., kitchen duties. Provisions for contributing to food resources through individual gardening activities provide sources of intrinsic and extrinsic rewards and group recognition. Combining station keeping greenhouse duties with small personal allowances for individual gardens has been found to be an effective motivational approach of combining work with pleasurable recreation.

Juri Gagarin was the first man to eat in space in 1961, followed by John Glenn a few months later. The first two-person-Gemini mission was also the first one where astronauts could enjoy a meal in space together. Luckily, they already had solid food, instead of food tubes. It was also the first time that astronauts brought fresh foods into this technical system albeit by breaking the rules (see Box 7.9).

Box 7.9 The First Sandwich Smuggled into Orbit

Just about two hours into the flight of Gemini 3, the first crewed mission on March 23, 1965, pilot John Young reached into his spacesuit's pocket and pulled out a surprise. "*Where did that come from*?" Gus Grissom, the mission's commander, asked his crewmate. "*I brought it with me*," Young replied, somewhat matter-of-factly. "*Let's see how it tastes. Smells, doesn't it*?"

What Young revealed was the world's first—and possibly last—corned beef sandwich to fly in space. The sandwich was a joke, although not everyone learning of it later would laugh. Grissom's fellow Mercury astronaut Wally Schirra, who had a reputation for pulling pranks (he called them "Gotchas"), had purchased the sandwich from Wolfie's Restaurant and Sandwich Shop at the Ramada Inn in Cocoa Beach two days earlier. He handed it off to Young the morning that Gemini 3 launched atop a Titan II rocket from Cape Canaveral's Pad 19 in Florida. "*It's breaking up. I am going to stick it in my*

(continued)

Box 7.9 (continued)
pocket," said Grissom after taking a bite. "*It was a thought, anyway… not a very good one*," Young replied, defeated. "*Pretty good though, if it would just hold together*." Crumbs were the concern. In the weightless environment of space, they could find their way behind electrical panels or fly into a crewmember's eye. […] Not everyone appreciated the joke. Some members of Congress were so upset that they investigated whether the contraband cuisine endangered the mission.

Source: Collectspace.com (2015)

On Skylab, astronauts had, for the first time, a large, dedicated area for food preparation and dining. They were eating together on a specially designed table, using knives, forks and spoons. From that point on, a table for having meals together has been considered important by the crew and, eventually, became a requirement for subsequent habitats. Although it is still difficult to have dinner together because of mission task scheduling, there is a wide agreement of its importance for the crew.

Eating daily main meal(s) in a group is a nearly universal behavior for a common social practice. Everyone in the world typically has at least one main meal a day, with different etiquette and social rules attached to each. These group meals are not only related to nutrition intake, but also reflect norms and rituals of a society. These serve as powerful, oftentimes, unconscious and unrecognized influences on expectations and behavior. Astronauts are no exception to taking their social norms with them. For example, on Skylab missions, observers noted that the crewmembers refused to travel over the table as they passed through the wardroom (Fig. 7.19a)

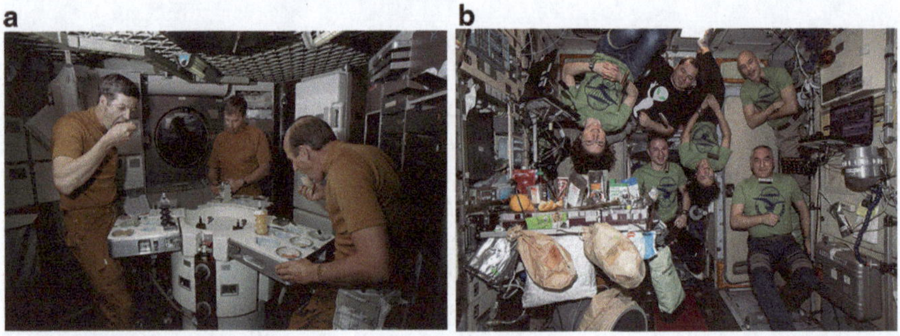

Fig. 7.19 (**a**) Eating together as designed. Skylab 2 astronauts eat space food in wardroom of Skylab trainer. The three members of the prime crew of the first human operated Skylab mission dine on specially prepared Skylab space food in the wardroom of the crew quarters of the Skylab Orbital Workshop (OWS) trainer during Skylab training at the Johnson Space Center. They are, left to right, Scientist-Astronaut Joseph P. Kerwin, science pilot; Astronaut Paul J. Weitz, pilot; and Astronaut Charles Conrad Jr., Commander (Credit: NASA). (**b**) Eating together onboard the ISS. ISS Expedition 61 crew in the Zvezda Service Module for a New Year's Eve meal on December 31st, 2019 (Credit: NASA and International Space Station Partners)

even though that was the most direct path. When asked about their circuitous path, they reflected and responded that it seemed inappropriate behavior to pass over the place where they eat. The Skylab crews unconsciously defined the space above the table as a 'sacred' zone and avoided it in the same way they would refrain from walking or sitting on a table on Earth. This unconscious defining of the space 'above' the table resulted in navigable space not being used in a microgravity environment. Pictures from the Mir space station as well as the ISS show that astronauts often strive to arrange themselves 'around' the table to fit everyone (Häuplik-Meusburger 2011).

Given the international nature of space missions, it will be a challenge to provide for multiple social practices and etiquette in a multi-cultural habitat. Although we can assume that practices will be aligned over time, the micro-society will be much more diverse than on Earth. This will require creative spatial solutions to avoid hindering social or individual mealtime activities.

7.7.4 Friendship, Intimacy and Sex

A defining characteristic of humans is that we are creatures of touch. We seek the company of others instinctively to find solstice and comfort. As primates, our childhood dependency on the protection of our families is prolonged engendering the bonds between members of different statures in our groups. Old with young, peer with peer, across genders, within genders, leaders to followers and vice a versa. Our complex affiliations define us and shape who we are. There will be friendships, intimacy and, oh yes, sex in space. Even astronauts are not immune to the emotions of attraction as was evidenced by the saga of Astronauts Nowak and Oefelein in 2007 (Sunseri 2007). When all missions were short and packed with activity, there was little time or energy to spare, as an astronaut put it: "*Sex is an interesting activity [...] If you find a good partner and you have time and you are not disturbed. [In space] there's no room, there is no time, there is no partner to do it*" (Haigneré 2009[23]).

But long duration missions offer both. And no amount of 'legislating' will prevent humans from being attracted to each other. When you only have a few others to interact with, there will be the struggle with over-familiarity. Being married will not be a panacea for sexual attraction to another. We will have to meet the matter frankly, train our crews to expect the probability, provide for natural affinities to be pursued and, as likely, ended. As our new societies evolve both culturally as well as societally, new forms of affiliations will emerge. Here, in this oh-so-sensitive subject of hearts and passions, will we have to show the greatest compassion and acceptance of our human natures.

---

[23]Haigneré, Jean-Pierre. 2009. Transcript not published. [interv.] Sandra Häuplik-Meusburger. Paris, France, June 2009.

In 1982 Svetlana Savitskaya, the second woman in space, was assigned a one-week visiting mission to the 211-day Salyut 7 mission (Shayler and Moule 2005). This was the first mission that included both men and women. According to Shayler, her five fellow male cosmonauts were shaving every day and tried to help her with her experiments. In return Savitskaya was cooking for them and the cosmonauts ate "*with renewed pleasure*" (Shayler and Moule 2005, p. 309). In 1991 the astronauts Jan Davis and Mark C. Lee were the first married couple to serve together on a mission (STS-47). They had married in secret because NASA regulations prohibited couples. However, intimacy and sexual activities are also possible without a partner. In his book 'Lift off' Michael Collins (1989) writes about the advice of a Skylab doctor to masturbate in order to prevent the possibility of getting infected prostate glands. The topic of 'sex in space' has been popular in science fiction (cf. Moonraker, Rendezvous with Rama by Arthur C. Clarke and others), but so far, the topic of intimate behavior has been avoided by national space agencies, and still is in simulation missions (Box 7.10).

Multiple studies have found positive impacts of mixed gendered teams and the normalization of group dynamics has universally been accepted as far outweighing risks of sexual attraction and interpersonal relationships. The inclusion of women has been found "*to relieve pressure and let normalize the culture of isolated groups*" in Biosphere 2 missions (Nelson et al. 2015, p. 81). According to Mark Nelson, one of the biospherians,[24] "*the tensions underlying actual and fantasized sexual liaisons or attractions—jealousy, intrigue—always powerful in human groupings, perhaps grow more so in isolated, confined groups*."

It's a difficult area to study. A study of mixed gender groups wintering over at the French polar station Dumont d'Urville in Antarctica (Rosnet et al. 2004) found the expected positive effects for the presence of women on the general climate of the group by reducing men's rude behavior. But Rosnet was able to discern some of the underlying factors that help explain ongoing dynamics. When participants were younger (20s–30s) and the females 'ages were close to the males', seduction behaviors appeared and rivalry, frustration and sexual harassment frequently resulted. These problems were less in evidence when the average age of both genders was either older or very different. When viewed within the context of human optimal mating ages, these findings make sense. It seems that the young males in the presence of young females will be most likely to engage in attempts to pair with available females and vice versa.

However, as space becomes more and more accessible, the variety of humans working and living in space will exponentially expand with the desired outcome of true settlement of extraterrestrial locations. Intimacy and sex are not a matter of age, but one of the 'right' conditions. With settlement will come families and children. It

[24]There were four women and four men in the crew. There were two couples who bonded before entry (and remain together 20 years after the closure experiment). Two individuals were in no sexual relationship, and two left partners outside.

is not a question of *if* but a question of *when*. So, we had better start designing for these fundamental human needs right now (Box 7.10).

Box 7.10 How to Turn a Single-Room into One for a Couple

The good thing was, that our rooms were already next to each other. But anyway, this probably would not have been a problem. . .We converted one bedroom into a lumber room, the other was the sleeping room. The beds were such that the frame was too narrow and uncomfortable for two people. We tried several options, but in the end, we moved my bed from my room over to his room—upright. Also, we put a lot of my clothes, which were also in boxes, in his room, together with his things. We practically had two mattresses next to each other in my room. In addition to these two mattresses, there was the table somewhere and this little drawer thing in the room. Then the room was already full. [. . .] I think it would have been better if you could redistribute or rearrange the rooms, or maybe change the shape of the rooms.

Additional storage space would also have been helpful. So according to the motto, I'll give you something from my corner here and I'll give you something from my side, or something like that. Just so that one has more freedom in the room design.

Source: Interview with a HI-SEAS crew member, 2018 (anonymous, not published)

7.7.5 Summary

Recommendations for habitat design that specifically address behavioral and performance challenges in ICEs include: the interior design of crew quarters, windows, opportunities for personalization, and social interaction. The practical environmental aspects of capsule co-habitation such as privacy, culture or territoriality have also been pointed out as critical in reviews of capsule environments. Those considerations have been included in the design of conceptual habitats (Bell and Hines 2005; Bishop 2006; Greene 2006; Petrov 2004). Recent works on human activity (cf. Haeuplik-Meusburger 2010, 2011) analyzed the spacecraft environment in relation to architectural programming, and established design directions for central human activities that make a 'home' in a long duration mission (especially deep space missions).

The goal is to incorporate features that can promote positive group interaction and reduce negative interaction. Desirable factors will promote group fusion, i.e., social bonding, group identification, social support. Negative factors to avoid are those that produce group fission, i.e., tension, conflict, discord, social isolation, scapegoating (blaming a convenient person), and miscommunication. Many factors can potentially contribute to either fission or fusion goals in multiple ways. For instance, separation of private functions from public is critical. This includes the provision of private spaces exclusively for one person, private spaces for a small group of people or a couple AND spaces where the whole community can come together. Separate spaces meet needs for solitude and privacy, allow individuals to regulate social interaction and provide for individual control over the amount of contact with others.

It provides for individual personalization and individual differences in activities while moderating feelings of crowding and confinement. In addition, there needs to be spaces that *require* and facilitate group interaction. One of the inherent dangers in isolated groups is the general drift of certain individuals into complete isolation from their crewmates (cf. Alexander et al. 1977).

The 'smart communication room' concept by Kalil Studio (1983–94)[25] is an early example for an architectural solution to multiple socio-spatial challenges. The 'floor' of this space would have been mechanized and computerized so that it would be able to unfold out of itself everything needed to work, exercise or relax. The idea was that "*the floor becomes the vibratory field of human interaction with space, adjusting to each need or desire. The entire room is developed into a responsive 'organism' capable of perceiving and responding to the needs and desires of individual*" (Kriegh et al. 2002).

Other innovative approaches embraced the fact that concepts like ceilings and floors are largely irrelevant in a microgravity environment. Kalil and Gardner (1983–1994) developed the idea of a community dish in which an undulating plane generally oriented along the axis of an enclosure and roughly dividing a cylindrical space into two sections with an opening through which access to both sides could be obtained. The pattern of waves built into the surface membrane conforms to the dimensions of the human body providing spaces reminiscent of natural features and spaces that provide individual and group community (Fig. 7.20).

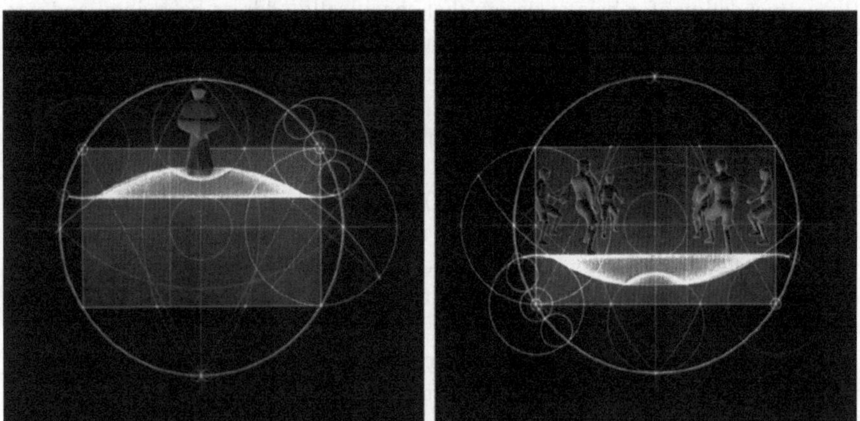

Fig. 7.20 Geometric Studies for the 'community dish' by Kalil Studio (as published in Kriegh and Gardner (2002), courtesy of Jean Gardner)

[25]During the years 1984 to 1986, the Kalil Studio engaged in two NASA sponsored contract cycles under the monitoring of Marc Cohen. The studio developed two preliminary design concepts to NASA's proposed habitation module.

Spaces that enforce human contact and interaction (e.g., meal and group leisure spaces) serve to preserve bonds between crewmembers as well as close contact necessary for accurate group monitoring and intervention when needed. On the other hand, when joined meal times are forced while experiencing social tensions within the crew, such a 'common rule' has the opposite effect.

To achieve these dual needs within the very limited space available in confined habitats, we need flexible, definable and re-definable interior environments that provide for group as well as individual activity.

7.8 Habitats for Humanity: Crafting Our Home Among the Stars

We are what we know. The argument can be made that truly new societies will not emerge until the old ones have completed their adaptation to the new conditions of extraterrestrial environments. The focus of the founding amalgamation of nationalities, cultures, values, and professions and individual beliefs, attitudes, norms, practices and expectations will be consumed with the process of adaptation, adjustment, accommodation, and assimilation. Beyond providing for the physical necessities for survival, the social necessities for wellbeing, community, belongingness, identity, and sense of place will be the work of the first generation and their children. Evolution may be constant but it begins with adaptation; and that sometimes takes time, trial, and error. We, must, of course, survive the trials and the errors.

What is needed is a paradigm shift that expands our understanding of built spaces and encompasses those characteristics that support wellbeing at the most fundamental level. Christopher Alexander[26] "cuts to the heart of the problem". In his book 'The Nature of Order' (2002) he notes that "*many of the [architectural] processes used today sadly are nearly bound to fail. We see the results of this failure all around us.*" (p. 3) "*Why don't we—why can't we—make places that feel beautiful and alive anymore*?" (Hora 2020). As highlighted by Dave Hora (2020), Alexander argues that it fails because our conception of order has become almost completely defined by models that describe the world in mechanistic terms. The success of the scientific method in explaining the mechanical aspects of the world has narrowed our viewpoint at the cost of being able to discuss the things that are truly important to us in the world, i.e., life of structure and human feeling. Because the mechanistic approach to explaining the world has been dominant, we, as a society have become convinced that mechanistic view is really all there is to the world, and that any value must come subjectively from us, and is not a part of the world around us. Without the existence of external value, "*we have no guide, no basis, and no common ground for producing a living world*" (Hora 2020).

[26]Christopher Alexander is a British-American architect and design theorist.

We were motivated to write this book because we believe a number of habitability factors can be deliberately employed to enhance coping abilities of those dealing with the extraordinary new living conditions posed by extraterrestrial environments. We have spent multiple decades professionally grappling with how to ensure the successful transition of humanity into space. Our individual careers have not only crossed paths upon multiple occasions, but we have had the opportunity to work alongside a multitude of wonderful, creative individuals from many, many professions. However, it has been a long journey as the focus slowly shifted from mere surviving to thriving. We believe we are finally at the threshold of a major paradigm shift where the living *space* becomes more important than the living *place*.

The future of habitability solutions is pregnant with possibilities! Enabling technologies involving the use of immersive mixed reality systems and haptic technologies to enhance work and leisure activities as well as provide enriched environmental stimuli are already on the near horizon. We know that current strategies used to ameliorate stress levels during short-term missions will be impossible during interplanetary voyages and Mars missions. For instance, support strategies that provide ISS crews private family real-time conferences and internet phone facilities will not be possible during a Mars mission. Therefore, telepresence and full fidelity audio/video/3-D communication replay capability will provide for more effective psychological support and interaction for crewmembers and more effective links to families and friends back on Earth.

The loss of real-time contact with family and friends will place a premium on selecting individuals who have effective coping strategies based on self-reliance and autonomy as well as a cooperative group focus. Although the burden of support will have to shift away from the 'absent' network of family and friends to members of the present group, crewmembers will also have to be able to balance needs for affiliation with autonomy and privacy needs. Both habitability and mission structure factors will play a significant role in generating group fusion (e.g., cohesion, teamwork, interpersonal trust) or group fission (e.g., discord, conflict, miscommunication, reduction in performance and wellbeing).

To maximize coping and adaptation, the living space for crews must actively contribute to the process. The 'best' solutions will be those that are passive, i.e., not requiring particular efforts on the part of inhabitants in order to benefit (e.g., one must use the treadmill before gaining any exercise benefit). Embedding the benefit by sheer presence or exposure ensures that inhabitants will be maximizing the potential for effective adaptation. Thus, incorporating facilitative features as part of the habitat seems to be a self-evident win.

A summary of structural layout and habitat design approaches discussed in this book includes but is not limited to:

- Maximizing habitable volume with configurations that are perceived as more spacious (e.g., physically separated and narrow spaces can be perceived to be unpleasant and cramped, versus an open floor plan with flexible partitions in relation to human activities can be perceived as pleasant).

- Utilizing multiple compartments, flexible partitions, adaptable furniture and alike to provide for variety and segregated use.
- Using color, lighting, texture and geometry to enhance desirable moods, reduce feelings of crowding and promote physiological synchronization.
- Using windows, digital displays, greenhouses or art to counter feelings of confinement and monotony and provide visual depth and to temporarily 'be somewhere else'.
- And designing for multiple uses of greenhouses for food production, leisure activities, stress reduction, crafts, gardening, small group interaction, exposure to full spectrum lighting and natural fractals.

This is not a book about finite answers. It is an invitation to the worldwide community of engineers, architects, psychologists, sociologists, human factors, and all others to collaborate, discuss, envision, dream, research, experiment and, ultimately, help craft our home among the stars. It will take us all to build our new world villages.

In order to contribute in the short term, a list of 'rules of thumb' has been prepared to summarize the analysis and review of terrestrial and extraterrestrial ICE habitats.

Ten Rules of Thumb for Habitability Design for ICE Environments

1. **One solution cannot fit all needs.**

   Individuals differ but there do seem to be consistencies in the personalities of those who fare well in ICE environments. It is more important to match the type of person needed to the type of mission.
2. **It takes time to uncover differences.**

   Every inhabitant brings his/her own cultural lens. Individual 'bias' may be completely unrecognized or subconscious. Differences in perceptions and subsequent interpretations lay the foundation for misunderstandings, miscommunications and interpersonal conflict as well as differences in coping and adaptation to the ICE environment.
3. **Long term missions have different requirements than short term missions.**

   Duration radically changes the importance of almost everything. Residing in place demands very different personalities than expeditionary missions that move from place to place.
4. **Everything you are *allowed to bring* must be a valuable resource.**

   One unbendable fact about the logistics of space travel and habitation therein is that everything is limited; physical space, storage, air, water, food, power, fuel, medicine ... everything, even people. This means that everything needed, especially in the beginning, must be brought. It also means that nothing that is not critical can be brought. We must approach habitability for extraterrestrial locations not only from the viewpoint of scarcity but one of finiteness.
5. **Integrated private, semi-private and social spaces are *necessary, not optional.***

   Being confined with the same group of people for long durations creates a need for more social space to enhance and limit social distancing over time, while also needing more private space to meet growing psychological needs.

However, creating additional space is limited; reconfiguring available space must be an option.

6. **Adequate volume is in relation to how space is perceived by the individual.** Too little (perceived) volume, behavioral constraints associated with a particular volume or too much chaos leads to feelings of 'crowdedness'. The most severe stressors produced by isolation and confinement involve monotony and boredom resulting from low workload, hypo-stimulation, restricted social contacts from family and friends, as well as a never-changing environment.
7. **Spaciousness can be increased by design and geometry.** The spatial composition of the interior in relation to human activities is a must in small volumes. Increase the spatial experience of the inhabitants by using textures, surfaces and also kinesthetic senses.
8. **Integrate natural environmental characteristics that support evolutionary derived perceptual processing.** It appears that not only is greenery restorative but we have been evolutionarily primed for optimal cognitive functioning in environments of particular botanical complexity. The complex properties of nature have been shown to be psychologically restorative due to its infinite variability, evolutionary association with beneficial resources, associations with relaxation and stress reduction activities.
9. **New environments offer new challenges and new possibilities.** Microgravity and reduced gravity environments not only present new challenges, they also afford new ways to view and use physical space. However, human perceptual systems may not accommodate all possibilities. Just because you can do it, doesn't mean it is an acceptable solution given human perceptual systems.
10. **Think transdisciplinary, be creative.**

It's not how large you make it, it's how you make it large.[27]

References

Abraham, F. D., Sprott, J. C., Mitina, O., Osorio, M., Dequito, E. A., & Pinili, J. M. (2003). *Judgments of time, aesthetics, and complexity as a function of the fractal dimension of images formed by chaotic attractors*. Retrieved May 20, 2020, from http://www.blueberry-brain.org/silliman/JEM%20ms2.htm

Adams, C., & McCurdy, M. (1999). Habitability as a tier one criterion in advanced space vehicle design: Part One-habitability (No. 1999-01-2137). *SAE Technical Paper* (1999). https://doi.org/10.4271/1999-01-2137.

Alexander, C. (2002). The nature of order. In *The process of creating life*. Berkeley, CA: The Center for Environmental Structure.

Alexander, C., et al. (1977). *A pattern language*. Oxford: Oxford University Press.

[27] Jim Wise, private communication, 2020.

Al-Sahhaf, N. A. (1987). *Contributions of windows and Isovists to the Juged spaciousness of simulated crew cabins*. thesis for the master of science in Engineering, University of Washington.

Associated Press (AP). (1997). Astronaut tells of down side to space life. *The New York Times*. Retrieved January 23, 2020, from https://www.nytimes.com/1997/01/22/us/astronaut-tells-of-down-side-to-space-life.html

Balling, J. D., & Falk, J. H. (1982). Development of visual preference for natural environments. *Environment and Behavior, 14*(1), 5–28.

Bell, L., & Hines, G. D. (2005). Mars habitat modules: Launch, scaling and functional design considerations. *Acta Astronautica, 57*(1), 48–58.

Bell, S. T., Brown, S. G., & Mitchell, T. (2019). What we know about team dynamics for long-distance space missions: A systematic review of analog research. *Frontiers in Psychology, 10*, 811.

Bishop, S. L. (2004). Evaluating teams in extreme environments: From issues to answers, international space life sciences working group (ISLSWG) group interactions workshop. *Journal of Aviation, Space and Environmental Medicine, 75*(7), C14–C21.

Bishop, S. L. (2006). Psychological and psychosocial health and well-being at Pole Station. In C. S. Cockell (Ed.), *Project Boreas: A Station for the Martian Geographic North Pole* (pp. 160–171). British Interplanetary Society.

Bishop, S. L. (2011). From earth analogs to space: Getting there from here. In D. Vakoch (Ed.), *Psychology of space exploration: Contemporary research in historical perspective* (1st ed., pp. 47–77). Washington, DC: NASA SP-2011-4411.

Bishop, S. L., Kobrick, R., Battler, M., & Binsted, K. (2009). FMARS 2007: Stress and coping in an arctic Mars simulation. *Acta Astronautica, 66*(9–10), 1353–1367.

Bluth, B. J., & Helppie, M. (1987). *Soviet space stations as analogs (NASA report NAGW-659)*. Washington, DC: National Aeronautics and Space Administration.

British Interplanetary Society. (2006). *Project Boreas*. Retrieved May 20, 2020, from https://www.bis-space.com/what-we-do/projects/project-boreas

Burrough, B. (1999). *Dragonfly: NASA and the crisis aboard Mir*. London: Fourth Estate.

Clearwater, Y. A., & Coss, R. G. (1991). Functional esthetics to enhance well-being in isolated and confined settings. In A. A. Harrison, Y. A. Clearwater, & C. P. McKay (Eds.), *From Antarctica to outer space* (pp. 331–348). New York: Springer.

Clervoy, J-F. (2009). [interv.] Sandra Häuplik-Meusburger. Transcript not published. Paris, France, June 2009.

CNN.com. (2000). *NASA Astronaut Jerry Linenger Recounts 'Adventure' Aboard Russian Space Station Mir in New Book 'Off the Planet'*. Aired Feb, 17, 2000–8:21. a.m. ET. Retrieved June 3, 2020, from http://www.cnn.com/TRANSCRIPTS/0002/17/ee.08.html

Cohen, M. (2009). Testing the celentano curve: An empirical survey of predictions for human spacecraft pressurized volume. *SAE International Journal of Aerospace, 1*(1), 107–142.

Collectspace.com. (2015). *Gemini 3 and the first (contraband) corned beef sandwich in space 50 years ago*. Retrieved March 9, 2020, from http://www.collectspace.com/news/news-032315a-gemini3-corned-beef-sandwich.html

Collins, M. (1989). *Liftoff: The story of America's adventure in space*. Grove Pr: New York.

Compton, D. W., & Benson, C. D. (1983). *Living and working in space: A history of Skylab. NASA history series*. Washington, DC: National Aeronautics and Space Administration.

Connors, M., Harrison, A. A., & Akins, F. R. (1985). *Living aloft: Human requirements for extended spaceflight*. Mountain View, CA: NASA Ames Research Center.

Craven, J. (2019). *The meaning of 'form follows function'*. ThoughtCo. Retrieved August 15, 2020, from https://www.thoughtco.com/form-follows-function-177237

Cummins, D. D., & Cummins, R. (1999). Biological preparedness and evolutionary explanation. *Cognition, 73*(3), 37–53.

Department of Defence. (2004). *Unified facilities criteria. Design: Arctic and Subarctic Construction – Buildings*. Retrieved May 14, 2020, from https://www.wbdg.org/FFC/DOD/UFC/INACTIVE/ufc_3_130_07_2004.pdf

European Commission (EC). (2020). *Housing space per person*. Retrieved July 20, 2020, from https://ec.europa.eu/energy/content/housing-space-person_en

Foucault, M. (1971). *The order of things*. New York: Vintage Books.

Geake, J. G. (1992). Fractal computer graphics as a stimulus for the enhancement of perceptual sensitivity to the natural environment. *Australian Journal of Environmental Education, 8*, 1–16.

Gifford, R. (2002). Making a difference: Some ways that environmental psychology has made a difference. In R. B. Bechtel & A. Churchman (Eds.), *Handbook of environmental psychology* (pp. 323–334). New York: Wiley.

Godwin, T. (1954). Cold equations. In J. Campbell (Ed.), *Astounding science fiction*, August. https://photos.state.gov/libraries/hochiminh/646441/vantt/The%20Cold%20Equations.pdf

Gotian, R. (2020). 7 astronauts offer tips on how to thrive while social distancing. *Forbes*. Retrieved August 14, 2020, from https://www.forbes.com/sites/ruthgotian/2020/05/21/7-astronauts-offer-tips-on-how-to-thrive-while-social-distancing/#1c028c5b40b2

Greene, M. (2006). Base design for pole station. In C. Cockell (Ed.), *Project Boreas: A station for the Martian geographic North pole* (pp. 32–48). London: British Interplanetary Society.

Griffin, B. N. (1978). *Design guide – The influence of zero-G and acceleration of the human factors of spacecraft design*. Lyndon B. Johnson Space Center, Houston, JSC 14581.

Hagerhall, C. M., Purcell, T., & Taylor, R. P. (2004). Fractal dimension of landscape silhouette as a predictor of landscape preference. *The Journal of Environmental Psychology, 24*(2), 247–255.

Hall, E. T. (1966). *The hidden dimension*. Garden City, NY: Doubleday.

Hall, L. (2015). *Setting the record: Fourteen months aboard Mir was dream Mission for Polyakov*. Rocketstem. Retrieved November 20, 19, from https://www.rocketstem.org/2015/02/09/russian-cosmonaut-valeri-polyakov-spent-record-breaking-14-months-aboard-mir-space-station-in-1990s/

Hartig, T., Evans, G. W., Jamner, L. D., Davis, D. S., & Garling, T. (2003). Tracking restoration in natural and urban field settings. *Journal of Environmental Psychology, 23*(2), 109–123.

Häuplik-Meusburger, S. (2010). My home is my spaceship – An investigation of extra-terrestrial architecture from a human perspective. In *40th International Conference on Environmental Systems*. AIAA 2010-6021.

Häuplik-Meusburger, S. (2011). *Architecture for astronauts: An activity-based approach*. Springer Science & Business Media.

Häuplik-Meusburger, S. (2014). Astronauts orbiting on their stomachs: The needs to design for the consumption and production of food in space. *Architectural Design, 84*(6), 114–117.

Häuplik-Meusburger, S., & Ozdemir, K. (2012). Deployable lunar habitation design. In V. Badescu (Ed.), *Moon: Prospective energy and material resources* (1st ed.). New York: Springer.

Häuplik-Meusburger, S., Aguzzi, M., & Peldszus, R. (2010a). A game for space. *Acta Astronautica, 66*(3–4), 605–609.

Häuplik-Meusburger, S., Peldszus, R., & Holzgethan, V. (2010b). Greenhouse design integration benefits for extended spaceflight. *Acta Astronautica, 68*(1–2), 85–90.

Heerwagen, J. H., & Orians, G. H. (1986). Adaptations to windowlessness: A study of the use of visual de'cor in windowed and windowless offices. *Environment and Behaviour, 18*(5), 623–639.

Hernandez, S., Brebbia, C. A., & De Wilde, W. P. (Eds.). (2010). *Eco-architecture III, Harmonization between architecture and nature*. London: WIT Press/Computational Mechanics.

Hietanen, J. K., & Korpela, K. M. (2004). Do both negative and positive environmental scenes elicit rapid affective processing? *Environment and Behavior, 36*(4), 558–577.

Hora, D. (2020). Retrieved from http://www.archnewsnow.com/features/Feature623.htm

Howe, S. A., Spexarth, G., Toups, L., Howard, R., Rudisill, M., & Dorsey, J. (2010). Constellation architecture team: Lunar outpost 'scenario 12.1' habitation concept. In *Proceedings of the Twelfth Biennial ASCE Aerospace Division International Conference on Engineering, Science, Construction, and Operations in Challenging Environments (Earth & Space 2010)*, Honolulu, Hawaii, March 14–17, 2010.

Joye, Y. (2007). Architectural lessons from environmental psychology: The case of Biophilic architecture. *Review of General Psychology, 11*(4), 305–328.

Kanas, N. (1971). *Behavioral, psychiatric, and sociological problems of long duration space Missions*, NASA TM C-58067, October 1971 National Aeronautics and Space Administration, Houston, TX.

Kanas, N., Sandal, G., Boyd, J. E., Gushin, V. I., Manzey, D., North, R., Leon, G. R., Suedfeld, P., Bishop, S., Fiedler, E. R., Inoue, N., Johannes, B., Kealey, D. J., Kraft, N., Matsuzaki, I., Musson, D., Palinkas, L. A., Salnitskiy, V. P., Sipes, W., Stuster, J., & Wang, J. (2009). Psychology and culture during long-duration space missions. *Acta Astronautica, 64*(7–8), 659–677.

Kaplan, S. (1987). Aesthetics, affect and cognition. *Environment and Behavior, 19*(1), 3–32.

Kaplan, S. (1988). Perception and landscape: Conceptions and misconceptions. In J. Nasar (Ed.), *Environmental aesthetics: Theory, research, and applications* (pp. 45–55). Cambridge: Cambridge University Press.

Kaplan, R., & Kaplan, S. (1989). *The experience of nature: A psychological perspective*. Cambridge: Cambridge University Press.

Kass, J., & Schiemann, J. D. (1996). Operational evaluation of the EXEMSI project. Experimental campaign for the European manned space infrastructure. *Advances in Space Biology and Medicine, 5*, 357–373.

Katcher, A., & Wilkins, G. (1993). Dialogue with animals: Its nature and culture. In S. R. Kellert & E. O. Wilson (Eds.), *The biophilia hypothesis* (pp. 173–197). Washington, DC: Island Press.

Kellert, S., Heerwagen, J., & Mador, M. (Eds.). (2008). *Biophilic design: The theory, science, and practice of bringing buildings to life*. Hoboken, NJ: Wiley.

Kennedy, K. J. (2009). The vernacular of space architecture. In A. S. Howe & B. Sherwood (Eds.), *Out of this world: The new field of space architecture* (pp. 7–21). Reston: American Institute of Aeronautics and Astronautics.

Klapp, O. E. (1986). *Overload and boredom: Essays on the quality of life in the information society*. Westport, CT: Greenwood Press.

Korpela, K. M., Klemettila, T., & Hietanen, J. K. (2002). Evidence for rapid affective evaluation of environmental scenes. *Environment and Behavior, 34*(5), 634–650.

Kotecki, P. (2018). Stunning photos show what daily life in Antarctic research stations is really like. *Business Insider*. Retrieved September 1, 2020, from https://www.businessinsider.com/daily-life-in-antarctic-research-stations-2018-11

Kriegh, M., & Gardner, J. (2002). *Kalil Studio: Proportion and meaning as key component of space station design*. AIAA Space Architecture Sympoisum, Houston Texas. AIAA 2002–6106.

Kwon, S., & Weed, N. C. (2019). *Neuroticism. Encyclopædia Britannica*. Retrieved May 14, 2020, from https://www.britannica.com/science/neuroticism

Lebedev, V. (1990). *Diary of a cosmonaut – 211 days in space*. New York: Bantam Air & Space Series.

Lefebvre, H. (1968). *Le droit á la ville*. Paris: Anthropos.

Lefebvre, H., & Busch, A. (2020). *Die Produktion des Raums: The production of space*. Spector Books OHG.

Leon, G. R., & Sandal, G. M. (2003). Women and couples in isolated extreme environments: Application for long-duration missions. *Acta Astronautica, 53*(4–10), 259–267.

Levi, D., & Kocher, S. (1999). Virtual nature: The future effects of information technology on our relationship to nature. *Environment and Behaviour, 31*(2), 203–206.

Linenger, J. (2000). *Off the planet – Surviving five perilous months aboard the Space Station Mir*. New York: McGraw-Hill.

Loria, K. (2016). What it's like inside the only permanent undersea research lab in the world. *Business Insider*. Retrieved October 14, 2019, from https://www.businessinsider.com/aquarius-reef-base-fiu-nasa-underwater-lab-2016-5

Mæland, B., & Brunstad, P. O. (2009). *Enduring military boredom: From 1750 to the present.* Basingstoke: Palgrave Macmillan.

Mann, A. (2017). *How will we deal with the psychological problems of spaceflight?* Retrieved August 28, 2020, from, https://medium.com/@adammann930/how-will-we-deal-with-the-psychological-problems-of-spaceflight-e191196006df

NASA [Bull.17]. (1975). Skylab Experience Bulletin No. 17: Neutral Body Posture in Zero-G. Houston, Texas, USA : NASA, Lyndon B. Johnson Space Center, 1975. JSC-09551.

NASA. (1978). *Anthropometric Source Book: Volume 1: Anthropometry for Designers, RP 1024* (pp. 7–78). Anthropology Staff/Webb Associates, NASA.

NASA [MSIS]. (1995). Man-Systems Integration Standards,Volume 1, Revision B,. July 1995. NASA-STD-3000.

NASA Analog Missions. (2018). *McMurdo Station*. Retrieved July 3, 2020, from https://www.nasa.gov/analogs/nsf/mcmurdo-station

NASA History. (2020). *Mir Space Station*. Retrieved March 2, 2020, from https://history.nasa.gov/SP-4225/mir/mir.htm

NASA Press Release. (2009). *NASA gives space station crew 'go' to drink recycled water*, 09-096, May 20, 2009. Retrieved August 2, 2020, from https://www.nasa.gov/home/hqnews/2009/may/HQ_09-096_Recycled_Water_Go.html

National Science Foundation (NSF). (2015). *McMurdo Station Master Plan 2.1*, December 16, 2015. Retrieved May 14, 2020, from https://www.usap.gov/news/documents/McMurdoMasterPlan_2.1.pdf

National Science Foundation (NSF). (2019). *Final comprehensive environmental evaluation for continuation and modernization of McMurdo Station area activities*. Retrieved July 23, 2020, from https://www.nsf.gov/geo/opp/antarct/treaty/cees/AIMS/Final%20CEE_McMurdo%20Modernization_v8_05Aug2019.pdf

Neighbour, P. M. (2001). *JAANUS Japanese architecture and art net user system.* Retrieved September 1, 2020, from http://www.aisf.or.jp/~jaanus/

Nelson, M., Gray, K., & Allen, J. P. (2015). Group dynamics challenges: Insights from biosphere 2 experiments. *Life Sciences in Space Research, 6*, 79–86.

Neuroscience. (2020). *Hikikomori: Understanding the people who choose to live in extreme isolation, Neurosciencenews*. Retrieved November 3, 2020, from https://neurosciencenews.com/hikikomori-social-isolation-17236/

Oosterhuis, K. (2012). Simply complex, toward a new kind of building. *Frontiers of Architectural Research, 1*(4), 411–420.

Organization for the Economic Cooperation and Development (OECD) Better Life Index. (2020). *United States*. Retrieved July 2, 2020, from http://www.oecdbetterlifeindex.org/countries/united-states/

Orians, G. H., & Heerwagen, J. H. (1992). Evolved responses to landscapes. In J. H. Barkow, L. Cosmides, & J. Tooby (Eds.), *The adapted mind. Evolutionary psychology and the generation of culture* (pp. 555–579). New York: Oxford University Press.

Palinkas, L., Gunderson, E. K. E., & Holland, A. W. (2000). Predictors of behavior and performance in extreme environments: The Antarctic space analogue program. *Aviation, Space, and Environmental Medicine, 71*, 619.

Pattara, L. (2016). *McMurdo Station, Antarctica: A mighty fine place to go for a drink.* Chivu Blog. Retrieved January 30, 2020, from https://www.chimuadventures.com/blog/2016/06/mcmurdo-station/

Pearlman, R. (Ed.) (2011). *CollectSPACE Interview: NASA's NEEMO XV aquanaut crew, posted 10-24-2011*. Retrieved March 9, 2020, from http://www.collectspace.com/ubb/Forum39/HTML/000292.html

Peldszus, R., Dalke, H., Pretlove, S., & Welch, C. (2014). The perfect boring situation – Addressing the experience of monotony during crewed deep space missions through habitability design. *Acta Astronautica, 94*(1), 262–276.

Petrov, G. I. (2004). *A permanent settlement on Mars: The first cut in the land of a new frontier.* Dissertation. Massachusetts Institute of Technology.

Porterfield, D. M., Neichitailo, G. S., Mashinski, A. L., & Musgrave, M. E. (2003). Spaceflight hardware for conducting plant growth experiments in space: The early years 1960–2000. *Advanced Space Research, 31*(1), 183–193.

Purcell, T., Peron, E., & Berto, R. (2001). Why do preferences differ between scene types? *Environment and Behavior, 33*(1), 93–106.

Quinn, S. (2009). Plants as important in space as on earth. Voanews. https://www.voanews.com/archive/plants-important-space-earth. Retrieved 15 May 2020.

realestatejapan.Blog. (2018). *How much living space does the average household have in Japan?* Retrieved May 20, 2020, from https://resources.realestate.co.jp/living/how-much-living-space-does-the-average-household-have-in-japan/

Rosnet, E., Le Scanff, C., & Sagal, M. S. (2000). How self-image and personality influence performance in an isolated environment. *Environment and Behavior, 32*(1), 18–31. https://doi.org/10.1177/00139160021972414.

Rosnet, E., Jurion, S., Cazes, G., & Bachelard, C. (2004). Mixed-gender groups: Coping strategies and factors of psychological adaptation in a polar environment. *Aviation, Space, and Environmental Medicine, 75*(7), C10–C13.

Sandal, G. (2002). Individual and group adaptation in space: Evidence from analogue environments. In *Proceedings of life in space for life on earth, 8th European symposium on life sciences research in space, 23rd annual international gravitational physiology meeting*, Karolinska Institute, Stockholm, Sweden, 2–7 June, 2002.

Sandal, G. (2004). Culture and tension during an international space station simulation: Results from SFINCSS'99. *Aviation Space and Environmental Medicine, 75*(7), C44–C51.

Santy, P. A., Holland, A. W., Looper, L., & Marcondes-North, R. (1993). Multicultural factors in the space environment: Results of an international shuttle crew debrief. *Aviation, Space and Environmental Medicine, 64*, 196–200.

Sauer, R. L., & Jorgensen, G. K. (1975). Biomedical results of Apollo. *Waste Management System.* Retrieved August 5, 2020, from https://history.nasa.gov/SP-368/s6ch2.htm

Schanche, D. (1959). The astronauts get their prodigious chariot. *Life Magazine, 47*(24), 112.

Schires, M. (2017). *Antarctic base McMurdo Station receives sustainable new master plan.* Retrieved August 3, 2020, from https://www.archdaily.com/878768/antarctic-base-mcmurdo-station-receives-sustainable-new-master-plan?ad_source=search&ad_medium=search_result_articles

Scopelliti, M., Carrus, G., & Bonaiuto, M. (2018). Is it really nature that restores people? A comparison with historical sites with high restorative potential. *Frontiers in Psychology, 9*, 2742.

Shayler, D., & Moule, I. A. (2005). *Women in space – Following Valentina.* Berlin: Springer-Praxis Books in Space Exploration.

Shields, J. (2017). *Scientists reveal what living and working in Antarctica is really like.* Retrieved January 30, 2020, from https://adventure.howstuffworks.com/scientists-living-working-in-antarctica.htm

Simon, M., Whitmire, A., & Otto, C. (2011). *Factors impacting habitable volume requirements: Results from the 2011 habitable volume workshop.* In D. Neubek (Ed.), NASA/TM-2011-217352. Center for Advanced Space Studies-Universities Space Research Association, Houston, April 18–21.

Sommer, R., & Summit, J. (1995). An exploratory study of preferred tree form. *Environment and Behavior, 27*(4), 540–557.

Spehar, B., Clifford, C. W. G., Newell, B., & Taylor, R. P. (2003). Universal aesthetic of fractals. *Computers & Graphics, 27*(5), 813–820.

Sunseri, G. (2007). Steamy E-Mails highlight NASA love triangle. *ABC News.* Retrieved September 3 2020, from https://abcnews.go.com/Technology/story?id=2926757&page=1

Synek, E., & Grammer, K. (1998). *Evolutionary aesthetics: Visual complexity and the development of human landscape preferences*. Vienna: University of Vienna. Retrieved May 15, 2020, from http://evolution.anthro.univie.ac.at/institutes/urbanethology/projects/urbanisation/landscapes/indexland.html

Taylor, R. P. (2001). Architect reaches for the clouds. *Nature, 410*.

Taylor, R. P. (2003). Fractal expressionism where art meets science. In J. Casti & A. Karlqvist (Eds.), *Art and complexity* (pp. 117–144). Amsterdam: Elsevier.

Taylor, R. P., Spehar, B., Wise, J. A., Clifford, C. W. G., Newell, B. R., & Martin, T. P. (2005). Perceptual and physiological responses to the visual complexity of Pollock's fractal dripped patterns. *Journal of Non-linear Dynamics, Psychology and Life Sciences, 9*, 89–114.

Ulrich, R. S. (1983). Aesthetic and affective response to natural environment. In I. Altman & J. F. Wohlwill (Eds.), *Human behavior and the environment* (Vol. 6, pp. 85–125). New York: Plenum Press.

United Nations Population Division (UN). (2000). *XII. Charting the progress of populations: Floor area per person*. Retrieved June 2, 2020, from https://www.un.org/en/development/desa/population/publications/pdf/trends/ProgressOfPopulations/14.pdf

Van den Berg, A. E. (2004). De charme van de Savanne: Onderzoek naar landschapsvoorkeuren [The charm of the savanna: Inquiry into landscape preferences]. *Topos*, 10–12.

Van Wijk, C. H. (2018). Coping during conventional submarine missions: Evidence of a third quarter phenomenon? *Journal of Human Performance in Extreme Environments, 14*(1), 12.

Walmsley, D. J., & Lewis, G. J. (2014). *People and environment: Behavioural approaches in human geography*. New York: Routledge Publisher.

WikiDiff. (2020). *Reside v. Live*. Retrieved January 30, 2020, from https://wikidiff.com/reside/live

Wilson, E. O. (1993). Biophilia and the conservation ethic. In S. R. Kellert & E. O. Wilson (Eds.), *The biophilia hypothesis* (pp. 31–41). Washington, DC: Island Press.

Wise, J. A. (1988a). *The quantitative modelling of human spatial habitability*. NASA-ARC technical report 177501, Moffett Field, CA.

Wise, J. A. (1988b). *The quantitative modelling of human spatial habitability*. Technical report (19890006159), Space Human Factors Office, NASA-ARC, Sunnyvale, CA.

Wise, J. A., & Rosenberg, E. (1986). *The effects of interior treatments on performance stress in three types of mental tasks*. Technical Report, Space Human Factors Office, NASA-ARC, Sunnyvale, CA.

Wise, J. A., & Taylor, R. P. (2002). *Fractal design strategies for enhancement of knowledge work environments*. Paper presented to the human factors and ergonomics society meeting, Baltimore MD, October 2002.

Wolfe, T. (1979). *The right stuff*. New York: Farrar, Straus, and Giroux.

Worden, K., Bullough, W. A., & Haywood, J. (2003). *Smart technologies*. London: World Scientific.

Zimmerman, R. (2003). *Leaving earth: Space stations, rival superpowers, and the quest for interplanetary travel*. Washington, DC: Joseph Henry Press.

Index

S. Häuplik-Meusburger, S. Bishop, *Space Habitats and Habitability*, Space and Society, https://doi.org/10.1007/978-3-030-69740-2